THE WOODLAND WAY

A Permaculture Approach
to
Sustainable Woodland Management

Ben Law

Published by

Permanent Publications
Hyden House Ltd
The Sustainability Centre
East Meon, Hampshire GU32 1HR, UK
Tel: 01730 823 311
Fax: 01730 823 322
Overseas: (international code +44 - 1730)
Email: info@permaculture.co.uk
Web: www.permanentpublications.co.uk

Distributed in the USA by:
Chelsea Green Publishing Company, PO Box 428, White River Junction, VT 05001
www.chelseagreen.com

First edition © 2001 Ben Law, new revised second edition 2013

The right of Ben Law to be identified as the author of this work has been asserted by him in accordance with the Copyrights, Designs and Patents Act 1998

Designed and typeset by Boyd Wardell and Tim Harland

All photographs by Ben Law unless otherwise stated

Cover photograph by Anthony Waters

Back cover photograph by Steve Morley

Printed and bound in the UK by Cambrian Printers, Aberystwyth

All paper from FSC certified mixed sources

The Forest Stewardship Council (FSC) is a non-profit international organisation established to promote the responsible management of the world's forests. Products carrying the FSC label are independently certified to assure consumers that they come from forests that are managed to meet the social, economic and ecological needs of present and future generations.

British Library Cataloguing-in-Publication Data

A catalogue record for this book is available from the British Library

ISBN 978 1 85623 127 5

All rights reserved. No part of this publication may be reproduced, stored in a retrieval system, rebound or transmitted in any form or by any means, electronic, mechanical, photocopying, recording or otherwise, without the prior permission of Hyden House Limited.

CONTENTS

Foreword ix
by Jean-Paul Jeanrenaud,
WWF International

Introduction xi
Our changing woodscape

Chapter One 1
Woodlands From The Wildwood To The 21st Century
Our woodland resource 3
The woodland ecosystem 3
Types of woodland 5

Chapter Two 19
The 21st Century & The Return Of The Forest Dweller
Charcoal burners 20
Sustainable rural livelihoods 22
Patterns of forest dwellers 28
The re-emerging woodland community 28
Low impact development 30
A case for the cabin? 31

Chapter Three 33
Woodland Assessment & Management Planning
Old records 34
Observation and recording 34
Assessing woodland flora 36
Access 36
A first assessment 37

Chapter Four 45
Establishing New Woodlands
Natural regeneration 46
Planting 46
Designing woodlands 47
Permaculture principles as 47
applied to designing woodlands
Two examples of designed woodlands:
Goudhurst 55
Meera's Wood 58
Planting trees and aftercare 59
Reclaiming degraded land 66
New planting initiatives 67

Chapter Five 69
Management of Woodlands
Establishing coppice 70
Buying a standing coppice 70
Coppice management 71
High forest management 79
Example:
Leckmelm Wood 83
Managing for wildlife 86

Chapter Six 91
From Tree To Finished Product
Cutting 92
Milling 92
Extraction 94
Using green wood 95
Seasoning wood 96
Example:
The Dunbeag Project 98
Adding value to coppiced wood 101
WoodFuel 104
Charcoal burning 105
Example:
The Bioregional 106
Development Group
Bodging 107
Steam bending 107
Weaving 108
Joinery 108
Example:
Prickly Nut Wood 109
Marketing of high forest produce 117
Certification 117
Sustainable development 119
in woodlands

Chapter Seven 121
Food From The Woods
The canopy layer 124
The lower tree layer 127
The shrub layer 129
The herbaceous layer 130
Ground cover 131
Climbing plants 132
Coppice fruit avenues 132
Brash hedges 134

Medicinal herbs	137
Fungi	137
Bees	139
Meat	140

Chapter Eight 143
Woodland Management & The Law

Felling licences	144
Tree preservation orders	145
Sites of Special Scientific Interest (SSSIs)	145
Insurance	146
Health and safety	146
Woodlands and taxation	147
Planning law and woodlands	147
Grants	155
Other Grants	155
Contracts	155

Chapter Nine 157
The Future

Localisation	160
Eleven woodland recommendations for the new millennium	160
Celebration	161

Epilogue 163

Appendices

1 Useful trees, planting conditions and produce	164
2 Trees for different conditions	169
3 Planning for a sustainable future	171
4 Relevant forestry planning cases	178
5 Sustainable rural livelihoods	179
6 Example of contract	180
7 Bibliography	181
8 Resources	185

Index 188

THE AUTHOR

Ben Law is arguably Britain's greatest living woodsman. He has inspired millions with numerous TV appearances, training courses and lectures, and was most notably voted by viewers of Channel 4's Grand Designs programme as having the Best Ever Build with his Woodland House. He has also been on the BBC programmes Countryfile and The Green Team. In America, Ben's profile is on the rise due to appearances on The World's Greenest Homes, The World's Most Extreme Homes (HGTV) and Nightline (ABC).

He has worked on a smallholding, growing fruit and vegetables and looking after livestock and, having gained an Advanced National Certificate in Agriculture, he became a shepherd and set up a conservation landscaping business, specialising in ponds and wild flower meadows. Woodlands were a natural progression and after seeking out a few experienced coppice workers, he began work in the woods and in associated coppice crafts.

In the late 1980s, Ben visited the Amazon looking for positive solutions to deforestation and on his return set up and directed the charity, The Forest Management Foundation, working primarily with community forestry. Ben was a founding member of the Forest Stewardship Council. He has also worked for Oxfam as a permaculture consultant.

Ben has lived and worked at Prickly Nut Wood for over 20 years, where he trains apprentices and runs woodland-based courses. He is most well-known, however, for his 10 year struggle to obtain planning permission to build a house in his own woods, made mainly out of materials from the woodland itself.

Ben is the author of five best selling books: *The Woodland Way*, *The Woodland House*, *The Woodland Year*, *Roundwood Timber Framing* and *Woodsman*. He is also the presenter of the *Roundwood Timber Framing* DVD, a film which acts as a companion to the book and records in detail one of his recent builds from beginning to end, as well visiting previous projects.

Ben is currently celebrating 30 years of working the land.

DEDICATION

This book is dedicated to my children
Rowan, Zed and Tess and to all children
who are fortunate enough to begin
their lives living in a woodland.

ACKNOWLEDGEMENTS

Finding time to write this book has been a personal challenge for someone who lives and works in the woods. I am therefore grateful to so many people who have helped me during this process. I would like to say a special thanks to Maddy and Tim Harland, friends and publishers. It seems astonishing to be back here again, revisiting *The Woodland Way* twelve years since it was first published in 2001. The cycle of life is truly mysterious, but always reliable. A special thanks to Brooke Medicine Eagle, on whose vision quest the message to write this book became so clear. To my brother Daniel for his support and our many memorable foraging sprees on the Devon coast. To Patrick Whitefield for giving so much time in reading and commenting on my original drafts. To Simon Fairlie, for his help with the planning section and his vast contribution to helping those struggling with the legalities of living on the land. To David Blair and Bernard Planterose for updates on their projects. To all those who have spent time working in the woods with me, thanks and keep coppicing. A special earthly thanks to Prickly Nut Wood, my home and place of growth and learning.

GLOSSARY

Bender: temporary home or shelter made from hazel branches and canvas covering, used commonly in woodlands.

Binders: twisted binding between stakes on a layed hedge, usually hazel – also called 'etherings'.

Bole: the main stem up to the first branch.

Brash: small branches from side and top of trees.

Burr: rough growth that develops on the tree.

Cant: a defined area of coppice, also regionally referred to as 'panel'.

Cleave: to split unsawn timber by forcing the fibres apart along its length.

Clump: small group of trees.

Coppice: broadleaf trees cut during the dormant season, which produce continual multi-stems that are harvested for wood products.

Coupe: a clearfelled area of woodland, sometimes coppice.

Crown: the branches and top of tree above the bole.

Faggot: a tied bundle of small branches traditionally used to fire ovens, now used for river bank restoration and coastal defence.

Glade: a clearing within the woodland.

Greenwood: freshly cut wood.

In-cycle coppice: coppice which has been cut at regular intervals and is not overstood.

Layering: pegging down a living stem to make it root and create another tree.

Lop and top: see brash.

Maiden: single stem tree that has not been coppiced.

Mast year: a year when trees produce a large amount of seed.

Overstood coppice: coppice that has not been cut for many years and is out of rotation for usual coppice produce.

Permaculture: an ecological design system to create a sustainable future.

Pollard: tree which has been cut above grazing animal height to allow repeated harvesting of poles from the crown.

Pleachers: partially cut through stems of hedges which are layed at an angle and continue to grow as the sap still flows.

Ride: an access route through a woodland often used for timber extraction.

Rootstock: the root onto which a scion is grafted.

Saw doctoring: sharpening of saw blades.

Scion: cuttings to be grafted onto a rootstock.

Shredding: management technique where side branches and top are removed from a living stem every 2-3 years for fodder and firewood.

Singling: converting a coppice stool to a single stemmed tree.

Snedding: removal of side branches and top of a felled tree.

Snigging: extraction of timber with horses.

Standard: a single stemmed tree allowed to grow to maturity, commonly amongst coppice.

Stool: the living stump of a coppiced tree from which new stems grow.

Stooling: the earthing up of a stool to ensure regrowing stems produce roots which can be cut and planted as new trees the following winter.

Subsoiler: deep cutting plough which helps break up compacted soil.

Suckering: regrowth from existing roots of a tree after cutting.

Sun shoots: fresh stems that grow from the base of overstood hazel coppice.

Swales: contour ditches that do not flow but collect water and help extra absorption into the soil.

Undercutting: the process of cutting the roots of a young tree in a nursery at a depth of usually 3-4in (75-100mm) below the surface. This stimulates root growth and removes the need to transplant.

Underwood: coppice woodland.

Whip: a young tree taken from the nursery to be planted out.

Yurt: a wooden framed transportable dwelling with canvas covering originating from Asia and now found more regularly as a dwelling in woodlands.

FOREWORD

Ben's brilliant book points the way forward for woodland management in the British Isles and beyond. It covers every aspect of what is in reality 'woodland stewardship', from both a practical and philosophical standpoint. Ben writes from the heart after long years of struggle with a whole host of naysayers who tried unsuccessfully to convince him by fair means and foul to give up his vision for a renaissance in the countryside. He has done what many have dreamed of doing: gone to live in his beloved Prickly Nut Wood in Lodsworth, and made a living from working close to nature. His book is a product of the profound wisdom he has gained from this experience and serves as a route map for others to follow. It is the best kind of read, both intensely personal and visionary, while at the same time immensely useful and full of handy hints for the many who share his vision. This book is set to be a classic and will surely mark a turning point in our relationship with woods and the natural environment.

Jean-Paul Jeanrenaud
Formerly Head of the Forests for Life Programme, WWF International
Currently Director of Corporate Relations, WWF International

EXPLANATION

Firstly, I want to mention the regular use of the word 'man' in this text as in woodsman. This is not a reference to gender, as men and women are equally capable of woodland activities. It comes from, *manus*, the Latin word meaning hand. Thus woodsman means 'hand of the woods'.

Secondly, I have chosen the use of 'woodland' over 'forest'. Historically, woodmanship was the skilled management of our ancient woodlands, and forest was a word used to describe the King's hunting ground, where he could hunt the deer. This consisted of large areas of both open ground and woodland. Woodland is a localised British word and, as localisation is an important element in this book, I have stuck with the word woodland in most of the text. Where I have used the word 'forest', it is used in reference to forest dwellers as this is an accepted global term for people who live, eat and earn their living from the forest.

INTRODUCTION

My interest in woodland stems from a love of trees, from a heartfelt need to protect and be close to these vast beings that stand so patiently in one place, observing, dancing in the breeze and carrying with them a wisdom of time and knowledge of place that we can only dream of understanding.

Like many people, the threat to trees and woodland from our fast pace of modern life and the encroaching tarmac and concrete, sent my life in the direction of 'looking out' for trees. I have longed for their protection, their freedom to grow old and gnarled, and as I have grown closer to trees and observed our relationship with them, it has become clear that as long as the human race also inhabits this planet, the future of trees is dependent on sustainable woodland management.

We need trees to clean the air we breathe, for wood for buildings and fire to cook on and as a place to share our stories around. We need the leaves for protein, the bark and flowers for medicines, the fibre for ropes and clothes, the sap for our wines, the fruit and nuts for our nutrition and above all their presence for our spiritual well-being.

In the British Isles since the last Ice Age, we have systematically eradicated wilderness in the bid to become civilised. We have cleared the wildwoods and moved further away from the symbiotic relationship of the forest dweller. In our attempts to control nature, we have built large and powerful machines capable of cutting, lifting and debarking a tree in one mechanical process. Yet whenever we stop for a moment, nature returns to reforest our striped lawns and uproot our highways. Her perseverance deserves our respect.

Modern forestry has degraded much of our landscape for the purpose of short-term financial gain. In this book, I look for alternatives to this pattern, taking knowledge from traditional practices, linked to our future needs. We can work with nature, harvesting in ways that cause minimum disturbance and sustain beautiful and productive woodlands for ourselves and other species.

I welcome the return of the British Forest Dweller.

Our Changing Woodscape

Over 12 years have passed, since I wrote the first edition of *The Woodland Way* and much has changed in the British woodscape. On the positive side, there has been a growing resurgence of 'woodlanders'; a new generation of land-based folk keen to manage and work the woods, sharing skills they learn and understanding the importance of a lifestyle that restores derelicts woods and begins to put in place a legacy for the next generation. Apprenticeships and training have been on the increase and, from my own perspective at Prickly Nut Wood, I have seen a number of willing people with limited skills evolve into confident woodsman who are actively working woodlands and finding a livelihood.

There has been a growth in education and awareness for all ages and the strong emergence of Forest Schools is enabling children today to engage with woodlands.

This in turn should help secure knowledge and appreciation of our woodlands within the next generation.

The proposed 'forest sell off' of large areas of Forestry Commission land into the private sector was abandoned in the face of strong public opinion and campaigning. It was wonderful to see the level of passion our woodlands generate when their future or fate is questioned.

The fate of our forests are at a critical period. The threat to a number of our tree species from plant pathogens has never been as high as it is today. The publicised news of *Chalara fraxinea* and its threat to ash trees is reminding many of the loss of our elm trees to Dutch elm disease at its peak in the 1970s. *Chalara* is one that has attracted publicity and the fact that the government has held two COBRA meetings shows this outbreak (although rather late) is being taken seriously. However, the arrival of Asian longhorn beetle in Kent, chestnut blight in Warwickshire and East Sussex, oak processionary moth in the Thames region, *Phytophora ramorum* devastating larch across the West Country and three other strains of *Phytphora* established, leaves scientists in fear of a *Phytphora* superbug mutation which could be less selective on species choice. With the number of these potential threats it is easy to become negative, but there is plenty we can do. Biosecurity needs to be increased at a local, national and international level. We need to tighten our controls on plant imports. It seems astonishing we considered importing ash when it so readily self seeds in the UK, but it is not just ash imports that are in question as the majority of our plant pathogens have arrived on imported plant material. We need to look at our timber imports and particularly focus on a ban of imported sawlogs. Bringing logs with bark into the country and all that resides beneath the bark needs to be stopped.

We need to extend biosecurity measures to ourselves as well. In Australia, you are sprayed upon entering the country and boots must be washed clean of mud. As an island we need to take a similar approach. At a local level, we can clean boots and machinery and not carry mud and plant material from wood to wood. As foresters we need to increase diversity in our planting. Only a richly diverse woodland is likely to be well positioned to protect itself from potential challenges ahead.

When considering diversity, we must plant more woodlands. Our forest cover in the UK is a poor – 13%, compared to EU cover of 37% – and a global cover of 30%. To increase our cover to 30% in the UK may seem a daunting task, but it is also an opportunity for employment in the planting and management of new forests, which in turn will help to ensure there is a plentiful supply of woodfuel and building materials for the next generation. The Woodland Trust is growing in members and influence and its work in looking after some of our ancient woodlands as well as embarking on the planting of the UK's largest new forest, Heartwood, again highlights the public care and concern for the future of our forests.

The planning system has been overhauled, and it is still too early to see the benefits or difficulties caused by the recently introduced National Planning Policy Framework. It seems so much will depend on the new local plans, of which in particular Pembrokeshire's Policy 52, the low impact planning policy, could make a useful template.

I hope that in reflection to the public's response assuring the Forestry Commission does not privatise woodlands, the Forestry Commission might consider some long-term leases to young people who wish to set up small-scale forest businesses and manage some of the neglected or currently unviable Forestry Commission woodland. These leases could be tied to a management plan to secure the long-term health and biodiversity of the woodland and retain public access throughout. This way livelihood opportunities would be created, neglected woodland would become managed and the public would not lose any access rights or 'ownership' over the woods.

Above all we must look after what we have, plant more, increase diversity and ensure the next generation has better local options for fuel, house building and their recreation than our generation is enjoying.

Ben Law
Prickly Nut Wood, 2013

WOODLANDS FROM THE WILDWOOD TO THE 21st CENTURY

Chapter One

In 1992, WWF published a book titled, *Forests In Trouble: A Review of the Status of Temperate Forests Worldwide*. In a case study of Britain it showed that its forest health is rated as one of the poorest in the European Community. Many trees in recorded surveys suffer from severe damage on the basis of crown density linked to air pollution from acid rain. As we rely on trees to filter and clean the air we breathe, the signs that they themselves are suffering have far reaching consequences for the human race and other species. Despite an increase in forest cover in Britain, the 20th century saw the quality of our woodlands decrease from an environmental perspective. Many of our richest woodland habitats have been neglected and protecting the biodiversity of the woodland has often been ignored in pursuit of short-term monetary gain. The result is an acidifying monoculture (grant aided by government) whose finished product, raw timber, is then tax exempt. Such policies are producing more timber (much of which ends up as pulp), but at a loss to our soils, our flora and fauna and the visual impact on our landscape. The incentive to continue this practice still remains and there is no flexibility with this plan. It is a case of plant, grow fast, harvest and sell. There is little or no consideration of non-timber products, people's recreation, or the forest workers' connection to place and the knowledge of woodmanship. Many of these forests are owned by insurance companies and pension schemes, and the interest is rarely in the forest but in the financial return of tax exempt timber sales and land speculation.

So where did it all go wrong? A brief journey through the history of our woodlands will shed some light.

The wildwood developed after the last Ice Age and slowly evolved not unlike a naturally regenerating woodland does today. Pollen analysis has shown that after the early pioneers like birch and sallow, climax species began to colonise localities. Lime was the main climax species in the south east of England with oak and hazel covering much of the rest of England and Wales, and Scots pine and birch in the Highlands. There were of course many other species and woodland types but this pollen analysis of about 5,000 BC gives us some idea of how this country looked before humans started to cut down trees and cultivate land. From this point onward the wildwood began to be cleared, (Source: *Trees and Woodland in the British Landscape* by Oliver Rackham) and it is estimated that by 500 BC half of England was no longer wildwood. As humans started to clear areas of wildwood, they became aware of processes like coppicing and suckering, and from this the conscious management of woodlands evolved. Grazing animals and fire were no doubt major tools in clearing the wildwood, creating wood pastures and eventually meadows.

As the centuries passed, the coppice woodland system became more complex in choice of species, soil types, and utilising the growth patterns of the underwood to shape standard trees above. By cutting hazel below oak standards the oak's shape could be altered to form curves for buildings or occasionally ships. Much of the coppiced wood was converted to charcoal, that formed the backbone of industrial fuel until the arrival of coal. Coal then started to be used as a preferred house fuel and when bricks replaced wood as the common building material the need for coppice woodland began to decline.

Many woodlands were cleared for agricultural land, some were left to be grown on into high forest, and others overplanted to form a plantation. It wasn't until the mid 20th century, however, that the quality of our woodlands really declined. The two World Wars used large amounts of timber but rarely destroyed whole woodlands, and at a time when food was short, the control of rabbits and deer, particularly through poaching, allowed natural regeneration and coppice regrowth.

The arrival of modern machinery and the increasing availability of the combustion engine had similar effects on woodland as combine harvesters have had on agriculture. Appropriate technology, that improves the efficiency of an operation without causing environmental damage, should be welcomed, but most postwar forestry equipment has caused environmental degradation. Extraction machines, unlike horses, have compacted and torn up woodland ride ecosystems that provided habitat to so many wild flowers and butterflies. Modern machinery has also yielded a modern workforce of forestry workers who operate in gangs, moving in and out of woodlands they do not know, cutting where they see a mark on a tree, having not even taken time to observe the woodland and its flora and fauna. This is a far cry from the relationship of the traditional woodland worker. In addition, planting patterns encouraged by grant aid have evolved into regimented rows of usually coniferous trees with straight forestry roads to allow in larger and larger machinery. Many of these plantations were established on ancient woodland sites and have slowly eroded the flora and fauna so unique to these particular woodlands.

Towards the end of the 20th century, we started to see a turnaround in our interpretation of woodland. An increasing urban population, the need for recreation in woodland and public pressures to preserve rare habitats have brought in some changes of policy. The Forestry Commission now offers a range of grants for improving biodiversity and public access to woodlands. Coppicing and traditional management techniques have started to return and native broadleafs are becoming the favoured planting species over conifers in England. The only danger of these positive outlooks is that in most cases they are monetary led. If the grants offered were removed how quickly might we fall back into the postwar patterns of forestry? Conservation and woodland management should be symbiotic without the need for a financial incentive to encourage us to look after our trees and other species. Our future must be to try to find ways to manage woodland sustainably, where all species, including ourselves, fulfil their needs from the woodlands without subsidies.

Our Woodland Resource

The British Isles has 6 million acres (2.4 million hectares) of woodland of which 3.8 million acres (1.5 million hectares) is coniferous high forest, 1.7 million acres (670,000 hectares) is broadleaf high forest and 49,000 acres (20,000 hectares) is coppice woodland with 520,000 acres (210,000 hectares) being non-productive woodland (Forestry Commission figures and areas at 31st March 1998). Of this woodland the Forestry Enterprise holds over 2 million acres (81,000 hectares) which are mainly in Scotland, Northern Ireland and Wales. This land is in theory the State's national forest. It belongs to us the people and we have freedom of access to it, although we have limited input as to how it is managed. In the Conservative government years from 1981 to 1987, 20 percent of the Forestry Enterprise land was sold, (about 400,000 acres/160,000 hectares) to private purchasers. Once it ended up in private hands, public access was removed or restricted in the majority of the woodlands. In 1998, the Labour government fortunately stopped the sales of Forestry Enterprise woodland. On a more positive note the Woodland Trust (Britain's leading charity dedicated solely to the protection of native woodland heritage) now owns about 50,000 acres (20,000 hectares) of woodland, all of which is open to public access.

Many of our woodlands are undermanaged, according to the National Small Woods Association. Out of 1.9 million acres (750,000 hectares) of broadleaf woodland in Britain, about 930,000 acres (375,000 hectares) are areas of woodland under 25 acres (10 hectares) in size, of which 430,000 acres (175,000 hectares) have received no formal management for at least 30 years. In some woodlands this may be a blessing, but as many are coppice woodlands which are overstood or plantations that were never thinned, the quality of the woodland is declining. This, of course, opens up an opportunity for rural employment, coppice restoration and small-scale sustainable woodland management.

The Woodland Ecosystem

The deciduous woodland is the most stable form of land use we have in these islands. The temperature within the woodland is far more stable than outside. In summer it is refreshingly a few degrees cooler as the evaporation by trees cools the air and, in winter, a degree or two warmer as night condensation of atmospheric water warms the air. The soils within the woodland are stabilised by the carpet of roots below the soil. The majority of roots are spread throughout the top soil layer, where they collect nutrients and water condensation of atmospheric water warms the air. The soils within the woodland are stabilised by the carpet of roots below the soil. The majority of roots are spread throughout the top soil layer, where they collect nutrients and water and interact with soil micro-organisms. Deeper roots stabilise the trees and tap nutrients and water held in the subsoil.

By contrast, when trees are felled in a clear felling operation, the root systems that hold the soil together begin to rot away and, when heavy rains or dry summer winds follow, the soil is washed or blown away and years of accumulated fertility is lost. The shade of the closed canopy ensures that soils do not dry out in the heat of the sun and the annual leaf fall constitutes a nutrient-rich mulch which helps to keep the forest floor moist. The crown and leaves of the tree work as interceptors, forming a baffle from heavy rain allowing the water to reach the soil at a gentle velocity. The rain also washes nutrients (bird manure) from the leaves to the woodland floor. Other nutrients are carried on the wind and the woodland intercepts them as well as acting as a shelter belt, protecting crops or settlements on the leeward side. Trees have the ability to reduce their leaf surface area to cope with high winds and this crown deformation by strong winds helps us to observe the direction and velocity of prevailing winds. On the convex slopes of hills and mountains, trees regulate the rush of run-off water that is baffled by the carpet of roots and built up layer of leaf fall. These help absorption, stop soils eroding and prevent the silting up of reservoirs and rivers.

One of the beautiful processes of a tree's life is transpiration. Trees and other plants release water through their pores which is returned to the air to make clouds. The clouds in return and interact with soil micro-organisms. Deeper roots stabilise the trees and tap nutrients and water held in the subsoil.

By contrast, when trees are felled in a clear felling operation, the root systems that hold the soil together begin to rot away and, when heavy rains or dry summer winds follow, the soil is washed or blown away and years of accumulated fertility is lost. The shade of the closed canopy ensures that soils do not dry out in the heat of the sun and the annual leaf fall constitutes a nutrient-rich mulch which helps to keep the forest floor moist. The crown and leaves of the tree work as interceptors, forming a baffle from heavy rain allowing the water to reach the soil at a gentle velocity. The rain also washes nutrients (bird manure) from the leaves to the woodland floor. Other nutrients are carried on the wind and the woodland intercepts them as well as acting as a shelter belt, protecting crops or settlements on the leeward side. Trees have the ability to reduce their leaf surface area to cope with high winds and this crown deformation by strong winds helps us to observe the direction and velocity of prevailing winds. On the convex slopes of hills and mountains, trees regulate the rush of run-off water that is baffled by the carpet of roots and built up layer of leaf fall. These help absorption, stop soils eroding and prevent the silting up of reservoirs and rivers.

One of the beautiful processes of a tree's life is transpiration. Trees and other plants release water through their pores which is returned to the air to make clouds. The clouds in return release the water as rain back to the trees. A large oak can transpire 80-120 gallons (364-545 litres) of water per day! In inland areas away from cloud formation by sea evaporation, removal of woodlands has the effect of decreasing the amount of rainfall.

The woodland itself is a complex assembly of symbiotic relationships. A squirrel plants a tree in the woodland by burying acorns for the winter harvest and not always returning to collect them all. The wood pigeon flies out beyond the woodland boundary and drops the seeds of the berry which has passed through its body. The seed germinates encased in a layer of manure. Thus the woodland spreads outwards absorbing the non-wooded landscape. Insects and fungi work to break down the leaves and dead wood creating the humus of the woodland floor. The alder absorbs nitrogen from the air and is able to fix the nitrogen in root nodules through a relationship with the Ascomycete fungus 'Frankia'. The nitrogen is then drawn up into the leaves and, when the leaves fall in autumn and then decompose, the nitrogen is made available to other species through soil enrichment.

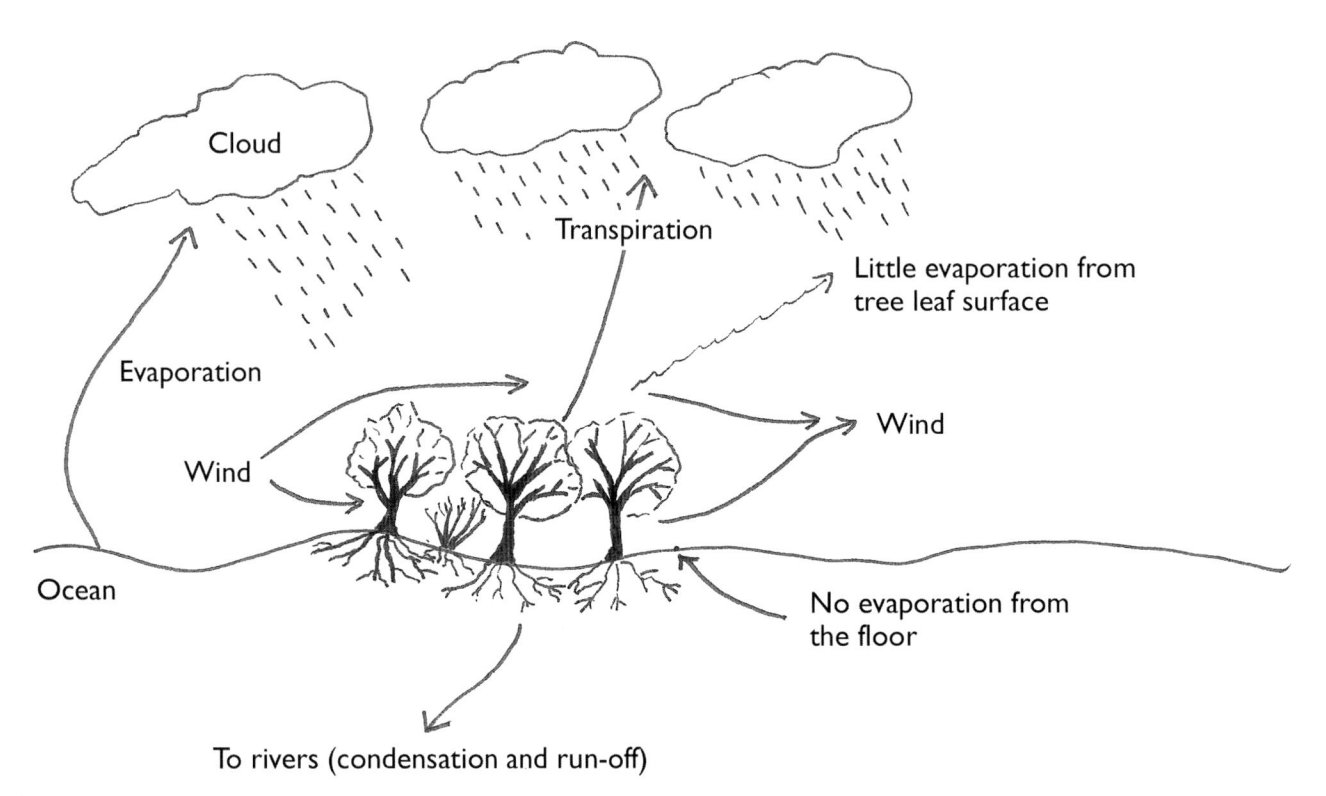

woodlands I would recommend visiting: Wistman's Wood on Dartmoor is a magical climax oak woodland, stunted and gnarled by the moorland winds and dripping with epiphytic ferns, looking as though they rise straight out of moss covered granite boulders; and Kingley Vale in Sussex is a climax yew woodland on the South Downs, dark and mysterious with burial mounds and breathtaking views.

To separate a tree from the complex relationship of the whole woodland ecosystem is to misunderstand the holistic patterning of nature. Our natural woodlands follow this pattern and when we start to create monocultural plantations, the web of life is broken and we create environments lacking in species, life force and fertility.

Types Of Woodland

Ancient Woodland

The term 'Ancient Woodland' is used to describe a woodland that is known to have been in existence for at least 400 years. The term 'Primary Woodland' is used to describe a woodland that has been in place since trees recolonised the British Isles after the last Ice Age. 'Secondary Woodland' applies to any woodland that has colonised open ground at any period after primary woodlands came into being, and may or may not be ancient. Ancient woodlands hold many clues to both the social and ecological history of the area and for us they are our greatest place of learning.

If I am visiting an area that I am unfamiliar with, I will search out an ancient woodland and spend time studying the variety of species and the history of their presence in the woodland. These will allow me to start to feel acquainted with the landscape of the area and empower me with a wealth of knowledge to translate into planting designs and species combinations. Above all ancient woodlands give me a sense of well-being and tranquillity and it is a privilege to spend time amongst ancient trees. These unique woodlands need our protection. There are two ancient

Opposite left:
Transpiration process.

Above:
Wistman's Wood, Dartmoor, Devon.

Below:
Climax yew forest, Kingly Vale, West Sussex.

Woodlands From The Wildwood To The 21st Century

The Caledonian Pinewoods

These ancient woodlands are the surviving remnants of what was the great Caledonian Forest which covered most of the Scottish Highlands. They have suffered greatly from over grazing by deer and sheep (and inappropriate planting of exotic conifers). The control of growing numbers of deer has brought the discussion of the reintroduction of wolves slightly closer to becoming a reality, although many hill farmers rearing sheep have a different viewpoint. Meanwhile many dedicated organisations like Reforesting Scotland and Trees For Life encourage preventative measures against deer and parts of the ancient woodland are slowly returning.

Coppice

Coppice is the term used to describe the successional cutting of broadleaf woodland during the dormant winter period. In spring, when the sap rises, the stump (known as the stool) sends up new shoots which are grown on for a number of years until they reach the desired size. They are then cut again during winter and the process repeats itself. The wood cut from coppice is known as underwood and has for centuries supplied a variety of traditional products and supported a large rural workforce, from the cutter to coppice merchant, craftsman and purchaser. Coppice is a valuable crop and managed well can sustain more people per acre than any of the modern forestry alternatives. It is also a sustainable pattern of management, rarely needing any replanting, so the soil is not disturbed and therefore not subject to the risk of erosion. Nutrients are returned mainly through the annual leaf fall. Coppice creates a cyclical habitat and unique ecosystem, and is one of the few patterns of symbiosis known in nature where humans are an important part of the relationship. In a well managed coppice, the stools are closely spaced, from about 4-6ft (1.2-1.8m) apart and the ground is fully shaded by the leaves and coppice shoots. When it is cut, sunlight pours in, dormant seeds waiting for light emerge and different birds, animals and insect life move into the newly created habitat.

Many rare species such as dormice and many types of butterflies are dependent on the coppicing system. My clearest experience of this comes with the migratory nightjar (normally a heathland bird, but with so much heathland disappearing it is now searching for alternative habitats). The male arrives from Africa in my woods usually during the last week of May, and sends its characteristic 'churring' call through the coppice, awaiting the arrival of a mate about a fortnight later. They nest amongst the stools of the freshly cut coppice. They follow my cutting pattern, settling in the cant I have last cut. If I did not continue the cycle of cutting, their habitat would disappear.

6   The Woodland Way

Coppicing is undergoing a revival and its value as an important landscape feature for social, ecological and commercial value is at last being seen. Traditional crafts are being revived and modern products made from coppice woodlands are finding niche markets, ensuring reasonable returns for those committed to this rewarding way of life. I would recommend a visit to Bradfield Woods in Suffolk, a coppice woodland managed since medieval times and now worked by the Suffolk Wildlife Trust.

Suckering

When you cut a broadleaf tree during the dormant season, the majority shoot as coppice but a few species do not regrow from the stump. Cherry, elm (except wych elm), white poplar and grey alder die at the stump but the roots stay alive and send up new shoots called suckers. These regrow and can be managed separately or in conjunction with coppice wood.

Coppice With Standards

The majority of coppice woodlands have some standard trees growing amongst the underwood. Standards make the optimum use of vertical space and give a high value return as well as the underwood produce. The key is to get the right balance of standards per hectare and to try and have a varied age of standards throughout the coppice. Too many standards will overshade the underwood and cause poor quality rods to grow on the coppice stools. One of the more traditional combinations is oak standards over hazel underwood. Oak and ash are the most common standards species but most species can be grown in this form, although I would exclude beech or hornbeam as they cast a heavy shade and little is able to grow beneath them. Most standards are developed first as maidens and then, after they have been cut, the stem is selected from the regrowth on the coppice stool and the others removed. This process is called 'singling'. It is a useful technique but the choice of stem must be central in the stool or it may over-balance due to wind blow.

Opposite top:
Caledonian pinewood forest at Rothiemurchus, Scotland.

Opposite left:
The coppicing cycle.

Above:
Mixed coppice at Bradfield Woods, Suffolk.

Right:
The suckering cycle.

Woodlands From The Wildwood To The 21st Century

Overstood, Stored or Neglected Coppice

When a coppice woodland is no longer cut on its regular rotation the rods from the stool continue to grow and the coppice becomes known as overstood. Sadly, in many parts of the country, this is the commonest form of coppice you are likely to see. An overstood coppice can often remain stable in its form for up to 40 years since the last cut, but from then on it will increasingly deteriorate year by year.

The root system of a coppice stool is used to support a specific height and weight of wood according to its rotation. When the coppice becomes overstood, the roots have to support a greater weight and height of wood and it becomes more susceptible to wind blow. Another common effect of overstood coppice is that as the woodland grows higher, stronger stools overshadow others and some begin to die out, therefore reducing the stocking rate of the wood. When the coppice cycle is stopped and the growth patterns of different species are no longer controlled by the levelling process of coppicing, diversity will decrease and some species will become dominant and overshadow others. For example, oak is capable of self singling and growing on into high forest. Overstood coppice can, however, usually be brought back into rotation (see Chapter Five).

Left:
Wild cherry grove (suckered), Lurgashall, West Sussex.

Top right:
Oak standard over sweet chestnut coppice, Prickly Nut Wood.

Below right:
Derelict mixed coppice, West Sussex.

8 The Woodland Way

Short Rotation Coppice

Short rotation coppice is as it sounds, coppicing on a rotation of usually one to four years. The most traditional short rotation coppice is basket-makers' willow (usually *Salix triandra* and *S. purpurea*), much of which is grown in the Somerset Levels. In parts of Sussex and Kent, sweet chestnut is cut on a two to three year cycle to produce walking sticks, the majority of which are sold to the National Health Service. Most short rotation coppice presently being planted consists of varieties of willow, poplar and alder which are grown as fuel wood. They are cut, chipped and then used to fire boilers to provide energy. This has been common practice for a number of years in Scandinavia and it is a positive renewable source of energy. The only negative aspect is that to obtain high yields, large inputs of fertiliser are often needed and the coppice needs to be dug up and replanted about every 25 years. However, as an alternative to monocultural cereal growing, and if the fertiliser can be of organic origin, it is a step in the right direction.

Pollarding

Pollard is the term used to describe a broadleaf tree that has been cut during the dormant season 6ft (1.8m) or more above the ground. The tree sends up shoots in a similar manner to a coppice stool but is cut at a height where animals cannot graze the regrowth of young shoots. The pollard poles are cut on a rotation and used in a similar way to coppiced wood. Pollards are often seen on wood banks (boundary banks subdividing or surrounding a wood), along river banks (usually willow cut on a short rotation for thatching spars), in towns (often limes or London planes), or in parkland (often oak or beech). An ancient pollard on a wood bank has historic interest as well as being a useful habitat, but from a management point of view climbing pollards to cut timber is labour intensive and not the safest of occupations. Beech is recommended as a pollard as it doesn't coppice as well as many other broadleafs.

Shredding

Shredding is the term used to describe a broadleaf tree that has had the side branches and sometimes the top removed leaving an expanding pole. The practice of shredding seems to have died out in England, but is still active in many other parts of Europe. I expect to see a revival in shredding, as it offers one of the most productive uses of trees, especially in the farm situation where fodder is needed for livestock. I spent time in the village of Vukaj in northern Albania where shredding is the most

Top:
Short rotation willow coppice used as part of the WETSystem (Wetland Ecosystem Treatment) at The Sustainability Centre, Hampshire.

Middle:
Ancient pollarded willows, Dedham, Suffolk.

Left:
Pollarded willow above Jacob sheep, silvi-pastoral system, Lodsworth, West Sussex.

Far left:
The pollarding cycle.

common woodland activity. The trees (in this case mainly oak) are shredded in late summer when there is still a good amount of protein in the leaves which are then stacked and stored to feed goats and cattle during the winter months. Cants are shredded on a two to three year cycle and any larger branches are used as cooking wood. The trees are healthy and clearly cope well with this type of management. Once the trunk of the tree has reached a useful size, it is felled and used for the construction of buildings. The trunks are fascinating pieces of timber, gnarled and notched from the continuous cutting of side branches and could be used to produce unique high value furniture in Britain.

From studying the shredded forests in Albania, together with my own experience of coppice management, I became aware of how shredding could become the early part of starting a coppicing cycle in Britain. With newly established woodlands, shredding could begin once the young trees are established (about eight years) and continue until the trunk has reached a useful size for furniture production (about 15 to 20 years). The trunks are cut during the dormant season and the woodland then moves from shredding to coppicing. The advantages are an increased yield from fodder and a high value product from the initial cut of the trunk which then starts the coppicing process.

Wood Pasture

Wood pasture is the term used to describe an area in which trees are undergrazed. The most well known wood pastures are wooded commons where some local people have commoners' rights to graze and collect wood from the common, although it is 'owned' by a landlord. Pollarding is a common practice in wood pasture. In Spain, examples of the *dehesa* system are still common

Top:
***Shredded oak forest,
Vukaj, northern Albania.***

Below:
***Oak woodland after cattle have been removed,
Lurgashall, West Sussex.***

Left:
The shredding cycle.

especially in Extremadura. Pasture is planted with holm oaks and cork oaks. The holm oaks are pollarded for charcoal, and the cork oaks stripped of their outer bark about every nine years to produce cork. The pasture is grazed by livestock and in the autumn fallen acorns are eaten by pigs which then produce the most delicious of hams, 'jamon iberico de bellota' (Iberian pigs fed on acorns).

Today we use the term agroforestry and there are many trials of different species as a tree canopy above grazing animals. Ash is a suitable species as it comes into leaf late allowing the pasture below to get off to a good start in spring. The shade is not too dense and, as the branches are cut in late summer (or the whole tree shredded) and left on the ground, the leaves which are high in protein are grazed and then the wood collected in the winter and burnt green. The following year more light will be allowed in as many branches will have been removed. Nitrogen fixing trees like alder and black locust will make a useful addition to any agroforestry system whether over pasture or arable crops.

Shelter Belts

Shelter belts can vary, from a narrow planted woodland to a rough hedge and, as their name suggests, they are planted to shelter specific features such as other crops, tender trees, buildings, livestock etc. Shelter belts are designed to allow some wind to permeate, but to take the main gusts of wind over the feature they are protecting. The shelter belts should be shaped so that the windward side is vertical and the leeward side slopes away creating more edge.

Plantations

The concept of plantations evolved to fulfil a need for 'quality planked timber'.

To build a house from roundwood poles is indeed a pleasure to the eye and to the builder, but it takes skill and joinery. The adze is a specialist tool and there are few amongst us who can shape a good and true plank from a cleaved log. A sawn plank from straight timber is still a prize for the forester both in economic terms as well as in its versatility of use. Plantations were and are needed to supply timber and the Forestry Commission was established in 1919 to ensure more timber was produced in Britain rather than imported from Scandinavia, America and the Tropics. This they did, and initially plantations were often single species planted in regimented blocks. The majority were coniferous and of little value to wildlife but they did produce timber! Pressure from conservation groups and the need for public access and recreation eventually encouraged the Forestry Commission to review its policy and the present situation is that they now offer larger grants to plant broadleafs rather than conifers, and even grants to restore overstood coppice. However, large-scale conifer plantations are still subsidised through planting grants and stronger enforcement is needed to deter large land-owning conglomerates from turning more of our land into a coniferous monoculture.

Opposite top:
Cork oaks in Alentejo, Portugal.

Opposite below:
Example of shelter belt at Keveral Farm, Cornwall.

Right:
Young oak plantation (not showing good forestry form).

Woodlands From The Wildwood To The 21st Century

Broadleaf Plantations

Broadleaf plantations tend to be densely planted and then go through a succession of thinnings and are eventually clear felled and replanted. The dense planting encourages straight growth as the stems are drawn up towards the light and the thinnings produce some income before the final clearfell, often up to 100 years after planting, which brings in the main economic return. Clearfelling always carries the risk of soil erosion and a move towards selective felling and restocking is a large step towards improving forestry plantations. (see Chapter Five).

In 1994, the Forestry Commission Bulletin No. 112, 'Creating New Native Woodlands', uses the National Vegetation Classification (NVC) of 19 types of woodland which are recognisable by their distinctive mixtures of trees, shrubs, flowering plants, ferns, mosses, liverworts and lichens. Each woodland type is limited to a climatic zone, soil type and expected climax vegetation which could be developed for the process of succession. The bulletin goes on to look at designing new native woodlands based on these classifications. At last we are looking at the different layers and relationships within the woodland environment and not just straight trees planted for plank timber.

Coniferous Plantations

Conifer plantations have been the dominant planting pattern of the past 60 years. Scotland, in particular has been subjected to vast plantations of sitka spruce, covering thousands of acres across the moorlands. The planting is purely economic as the return on these plantations comes within the lifetime of the landowner, although much of the timber goes to pulp as it is white in colour. Little is good for saw logs because in Britain, compared with Scandinavia, it grows too quickly (the rings become spaced out, resulting in poor performance in stress testing). This is an attitude of short-term thinking for financial gain – the damage to the soil through acidification will be left to the next generation to clean up. Experimental plantations of mixed conifers planted in blocks can also be found. Many in the highlands have now come to maturity but the economics of cutting and extraction makes the process uneconomic. Consequently plantations are being sold off at low prices. The economics are based on felling gangs, machinery and the sale of much of the timber for pulp, and these economics are crippling much of the forestry industry.

When we look for alternatives – such as adding value on site, mobile milling and local marketing – these plantations have value and should be used. When we look at monocultural coniferous plantations, as opposed to a mixed broadleaf woodland it sometimes seems difficult to understand why so many have been planted. Looking at the vast plantings in the Highlands, much of the planted land is relatively infertile with regard to growing commercial broadleaf varieties and so such species as sitka spruce are often seen as the most suitable choice. Also most modern foresters who have been through a forestry education at college are often taught to believe that only conifers are profitable on such land, (again the short-term economic thinking model) hence creating a culture of pro-coniferous foresters. Yet species like birch, regarded as a 'weed species' by many modern foresters, is a native tree that grows well on the poor highland soils. It has many possible markets, from the more traditional markets for besom brooms, horse jumps, faggots and firewood through to turnery wood, furniture making, toys, charcoal, floor boards, bark veneer and fire lighting strips. This is to say nothing of its non-timber products such as its wine making sap, mushroom cultivation, material dyeing qualities and the nitrogen enriching and pH enhancing qualities of its leaf litter.

Where we have large established plantations, we need to look for ways to bring about transformation from the vast monocultural woodland towards the more multifunctional woodland. Grizedale Forest, in the Lake District was a typical coniferous plantation but through the opening up of parts of the plantation and incorporating walks, a sculpture trail, artist in residence schemes and the 'Theatre in the Forest' as well as planting broadleafs, the dark woods are opening up to the human need for recreation. Large plantations can and should be diversified: a selection of clearings and glades can be created, interesting walkways, sculptures and the conversion of blocks or strips to broad-leaf can create the rolling change over a generation from a dark monoculture to a place of interest and, in the long-term, a more diverse and ecologically valuable woodland. This is a woodland that produces timber, supports wildlife, provides habitats for rare species and gives people a place to relax.

Opposite left:
'The Ancient Forester' by David Kemp, Grizedale Forest.

Above:
Douglas fir plantation, Lickfold, West Sussex.

Mixed Plantations

Mixed plantations are usually planted with the long-term aim of high forest broadleaf. The coniferous nurse crop is planted with the broadleafs, and the broadleafs are drawn up by the darkness around them cast by the faster growing conifers. The conifers are thinned out early giving some financial return, leaving the now straight broadleafs to mature. Usually they are planted in straight rows often three broadleafed rows then three coniferous etc. These mixtures often benefit the final timber provided thinning is carried out at the right time. Well-tried combinations are beech nursed by Corsican or Scots pine and oak nursed by European larch or Norway spruce. Broadleafs can also benefit from being planted in established gorse (provided the gorse does not overshadow the broadleafs) which gives shelter on frosty sites and acts in a similar way as a nurse crop. In pure broadleaf plantations, alder has been used on poor soils as a good nitrogen fixing nurse to ash.

Orchards

Orchards in the traditional sense could come under the heading of wood pasture. A traditional orchard was once part of almost every farm. The trees were large standards beneath which there was grazing pasture. My first job was on a smallholding with a traditional orchard in decline. Under the fruit trees, we ran sheep, and ducks, geese and hens free ranged. They all benefited from the windfalls, manured the orchard, grew well on the old pasture and no doubt feasted on many of the pests that would otherwise have affected the fruits. In a good year, the fruit was abundant and good cider and jams were a regular part of life. Those old orchards were planted with at most 50 trees per acre whilst a modern orchard has maybe 300 dwarf rootstock trees to the acre and two or three varieties at best. To maintain these densities, large amounts of fertilisers and pesticides are usually applied. These new orchards are of little ecological value compared to traditional orchards which are sadly becoming harder to find as many have been grubbed out due to insensitive grants from the former Ministry of Agriculture, Fisheries and Food (MAFF). In losing our orchards we are losing an important part of our cultural heritage and many of the local varieties that gave different parts of the country a sense of identity. We can replant orchards and grow local varieties by collecting cuttings from old existing trees to create places of recreation and abundance once again. To add to the absurdity of subsidised land use, the MAFF grants which paid for the grubbing out of orchards are now replaced by Countryside Agency grants to replant them!

Continuous Cover Forestry

As we search for an alternative to clearfell or a different approach to a 'shelterwood' system of management (see Chapter Five), turning to continuous cover forestry, or 'irregular forest management' as it is often referred to in France, offers the best option for diversity and age of species, along with protection of forest soils. Managing high forest to create a complex mixture of species and ages with a vigorous underwood and dappled shade for natural regeneration whilst still producing quality timber is an ideal that at last more foresters are turning too. The excellent charity Woodland Heritage has created a number of Continuous Cover Forestry Groups (CCFG) to study and monitor trials for converting forests and for new plantings of continuous cover forests.

Hedgerows

Like the old orchards, our hedgerows have also suffered a vast decline. Mechanisation and the growth of farm machinery climaxing in the arrival of the vast and cumbersome combine harvester (too large to turn and operate in small hedged fields) has been devastating for our hedgerows. Between 1945-1970, 140,000 miles (225,000km) of hedgerow were lost (Pollard, Hooper and Moore, 1974) but, like coppicing in many parts of the country, the revival of hedgerows and hedgelaying is on the increase. Hedgerows have many important functions. They provide a windbreak for crops and livestock, they provide firewood, wild fruit and nuts, a habitat for many species and a wildlife corridor linking woodlands across otherwise open arable areas. They help prevent soil erosion from wind, and if looked after and layed, they are a permanent feature, visually pleasing on the landscape.

Hedgelaying is the process where a hedge is cut and laid at an angle (about 30 degrees uphill) during the dormant season. The cut stems known as 'pleachers' are not cut right through; bark, bast, cambium and a small amount of sap wood are left. This allows the sap to flow through the pleachers which send out vertical shoots. Stakes and binders are used in some hedgelaying styles to secure the pleachers and leave a visually attractive hedge. Regularity of laying depends on the choice of species, the climate and how often the hedge is trimmed. About every 15 years for an untrimmed hedge is a good estimate. A trimmed hedge should be thinned and then left to grow on for about seven years before laying. A well laid hedge will be stock proof, removing the need for erecting and maintaining fencing. The planting and restoring of hedgerows is now grant-aided by local councils and the Countryside Agency. Old hedgerows often contain standards or pollarded trees. These provide a source of firewood, wild food and timber, but above all add an extra layer for wildlife benefit and visual beauty. Standards and the creation of pollards should be encouraged in new hedgerow plantings. The planting of nut bearing trees such as walnuts or sweet chestnut would add extra yield as standards. Ash would be a good choice as a pollard, breaking into leaf late and supplying a good source of firewood from the pollarded poles. Some legal support for the security of hedgerows came just before the fall of the Conservative government in 1997 with the Hedgerow Act, which means that landowners and farmers must apply to their local authority to get permission if they wish to remove a hedge. (Department of Environment Hedgerows Regulation 1997; statutory instruments no.1160; HMSO, 1997). Let us hope that such a scheme could be extended to include other landscape features under pressure such as the draining of wetlands or afforestation of moorlands.

Opposite left:
Nurse crop planting of Lawson cypress and beech. Aftercare and thinning has not been carried out.

Top:
Old Orchard, Lickfold, West Sussex.

Below:
Layed hedge, Midland style.

THE 21st CENTURY & THE RETURN OF THE FOREST DWELLER

Chapter Two

The Charcoal Burner

The Charcoal Burner has tales to tell.
He lives in the forest,
alone in the forest;
he sits in the forest,
alone in the forest.
When the sun comes slanting between
the trees,
and rabbits come up, and they give him
good morning,
and rabbits come up and say,
'beautiful morning'...
When the moon swings clear of the tall
black trees
and owls fly over and wish him
goodnight,
quietly over to wish him goodnight ....

And he sits and thinks of the things
they know,
he and the forest, alone together –
the springs that come and the
summers that go,
autumn dew on bracken and heather,
the drip of the forest beneath
the snow ...

All the things they have seen,
all the things they have heard:
an April sky swept clean and the song
of a bird ...
oh, the Charcoal Burner has tales
to tell!
and he lives in the forest and knows
us well.

by A.A. Milne, 1927

From *Now We Are Six* © A.A. Milne. Copyright under the Berne Convention. Published by Methuen, an imprint of Egmont Children's Books Ltd., London and used with permission.

'The Charcoal Burner', from *Now We Are Six* by A.A. Milne, illustrated by E.H. Shepard, copyright 1927 by E.P. Dutton, renewed (c) 1955 by A.A. Milne. Used by permission of Dutton Children's Books, an imprint of Penguin Putnam Books for Young Readers, a division of Penguin Putnam Inc.

Line drawing by E.H. Shepard copyright under the Berne Convention, reproduced by permission of Curtis Brown Ltd., London.

Charcoal Burners

The focus of this book is a way of life which I call 'The Woodland Way'. It is a way of living where humans are integrated as fully as possible into the ecosystem in which they dwell. This is not some romantic vision of a Utopia; it is based upon the inputs and outputs of a way of life that has been practically tried and tested by generations of forest dwellers across the globe.

For centuries woodland workers have lived and worked the woods of these islands. From common lands to large estates, woodlands were filled with craftsmen working the underwood and converting it into finished saleable products. As Milne's poem clearly shows, in 1927 it was a common site to meet a charcoal burner living in the woods. The history of the demise of charcoal burners and their more recent reappearance is market led. Charcoal burners and their families have lived and worked in woods as long as can be remembered. They lived in traditional huts, usually conical in shape, often grouped in hut circles.

They were wooden framed, lined with sacking and usually covered with turf, although some with a stone walled base have been found in Cumbria. Charcoal burners continued their lifestyle and craft, most likely unaware of planning legislation growing more complex around them. After the Second World War, charcoal burning diminished with forges using cheaper coke to heat metal and fossil fuels removing much of the need for traditional charcoal. But, with the interest in barbecues in the late 1970s and early '80s, one that is still expanding by about 5 percent per annum, the need for charcoal returned. Most charcoal available was and still is imported, often coming from fragile environments like the Indonesian rainforests and from mangrove swamps.

The growing demand for locally produced charcoal (that can be made from restoring overstood coppice) is reviving the charcoal burner's return to the woods. However, their need to live with their work – which has been unquestioned for thousands of years – is now often objected to by local authorities.

In Britain, the forest dweller has almost been hunted to extinction by bureaucracy and the pace of development. But now we are reaching a unique

Top:
Charcoal burners huts, Chase Lane, Haslemere, Surrey/Sussex border.

Below:
The author's first charcoal burning dwelling – a hazel structure with canvas covering, known as a bender.

position in our evolution where our understanding of our role within the ecosystem and the importance of being integrated rather than separated from nature becomes self-evident, and consequently more people are returning to the woods.

Sustainable Rural Livelihoods

Returning to live in the woods in the 21st century may seem to some people to be a radical approach to woodland management. When we look at concepts of sustainability and patterns of global forest dwellers, in the case of the diverse woodsman it seems the most sensible way forward.

So why is it necessary to dwell in the woods?

Transport and travel are key concepts of sustainability. To travel to and from the workplace is an unnecessary and polluting use of energy. Traditionally, those working continuously on the land, whether farmers or foresters (as in the case of charcoal burners), have resided where they work, to look after the land and pursue the activities carried out on it.

At Prickly Nut Wood, where I live, the need to be in the woods to control the burning of charcoal at night is one such example. If I was to live near my woodland rather than within it, I would have to travel daily to and from work and at regular intervals during the night to check the fire in the kilns. In my local village, houses often sell by unusual methods. Estate agents have names of people who are looking for houses in the area, and the demand is so high that the houses are regularly bought over the telephone before the new owners have set eyes on the property. An average small cottage sells for £400,000 and above. The inflation of house prices in

Above:
Charcoal kiln burning at Prickly Nut Wood.

Above Right:
Woodlanders' huts in Cumbria.

Greenwood Trust

my area has grown at a rate far in excess of the market value of the increase of coppice products. The woods surround the village and need to be worked, but the cost of finding housing is far beyond the means of the coppice worker. The need for on-site accommodation is the only opportunity for me and for many others in similar circumstances.

Other key reasons for residing in the woodland are the importance of gaining an intimate knowledge of the woodland and its flora and fauna. At Prickly Nut Wood, I know where the deer rest up during the day, and by being there I am able to take an active part in ensuring protection of coppice regrowth from browsing deer. I add value to produce in the cants of freshly cut coppice, keeping up activity and leaving human scent while the young shoots of coppice regrowth are at their most vulnerable from deer grazing. I know the badger sets and their night time pathways. I know where the early purple orchid, *Orchis mascula*, appears and I am careful to avoid working in that area from March to June. This knowledge is fundamental to all forest dwellers, and is often absent in modern forestry.

The produce from Prickly Nut Wood is diverse. Non-timber products as well as the coppice wood products are harvested throughout the year. The birches and maples are tapped for their saps to make wine. The fruit trees are pruned and a vigilant watch is kept during early spring to ensure fruit buds are safe from bullfinch attack. The harvesting of fruit, nuts, berries and leaves are regular summer and autumn activities. I keep bees, and when they swarm, I hear them and can follow; I can catch the swarms and increase my number of colonies because I am there, living on and from the land. I am there to harvest fungi, which can be spoilt for the palate if left too long unobserved, and the regular tending of my raised vegetable beds are all part of a management system designed by the forest dweller.

Another key reason for dwelling within the woodland is the sense of locality. My produce sells locally to local people. This cuts down on the vast distances timber is transported unsustainably around the country and from beyond. My local identity and the relationship that I have with the local community stems from people knowing me as 'Ben the woodsman', who dwells in the woodland.

Sadly, security is also an issue in the 21st century. The need to dwell within the woodland to safeguard tools and machinery is now also an important need.

Seasonal calendar of the diversity of activities carried out at Prickly Nut Wood

	Jan	Feb	March	April	May	June	July	Aug	Sept	Oct	Nov	Dec
Bees			+	+	+	+	+	+	+			
Bracken cutting				+	+		+					
Bullfinch watch			+	+								
Charcoal	+	+	+	+	+	+	+	+	+	+	+	+
Cut coppice	+	+	+							+	+	+
Deer watch				+	+	+	+	+	+			
Faggott making	+	+	+	+						+	+	+
Firewood	+	+	+	+						+	+	+
Fruit harvest					+	+	+	+	+	+	+	
Furniture				+	+	+	+	+	+			
Mushrooms			+	+					+	+	+	+
Ponds				+								
Pruning	+	+						+				+
Rhododendron control				+	+	+	+					
Ride management		+		+	+		+	+				
Saps		+	+									
Security	+	+	+	+	+	+	+	+	+	+	+	+
Vegetable growing			+	+	+	+	+	+	+	+		
Walking sticks		+	+	+	+							
Wildfood	+	+	+	+	+	+	+	+	+	+	+	+
Yurt making				+	+	+	+					

The Woodland Way

I see my livelihood at Prickly Nut Wood as an example of what the British government calls 'sustainable rural livelihoods'.

> "The 1997 U.K. government's White Paper on international development commits the Department for International Development (DFID) to promoting 'sustainable livelihoods' and to protecting and improving the management of 'the natural and physical environment.'"
> From *Sustainable Rural Livelihoods*, edited by Diana Carney, DFID, 1998.

If the British government is committed to promoting sustainable rural livelihoods internationally, I presume they are equally committed to promoting them in Britain.

> "A livelihood comprises the capabilities, assets (including both material and social resources) and activities required for a means of living. A livelihood is sustainable when it can cope with and recover from stresses and shocks and maintain or enhance its capabilities and assets both now and in the future, while not undermining the natural resource base."

The key framework DFID uses is a pentagon of capital assets, that enable livelihoods to be analysed from a more holistic viewpoint. The pentagon in the box below is a useful model for looking at the sustainability of the livelihood of the forest dweller. If I take my own livelihood based as a forest dweller at Prickly Nut Wood, I can fill in the areas of capital assets as follows:

It is clear that my livelihood at Prickly Nut Wood has a good balance of different capital assets and therefore should be diverse enough to survive difficult times and look after the natural resource base of the woodland while living a sustainable rural livelihood. It clearly shows that my livelihood as a forest dweller has roots in a certain piece of woodland. It has interaction with a particular locality for sales of produce, help with labour when needed and social support through relationships of trust.

Groups such as the Sussex and Surrey Coppice Group are forums for debate on relevant issues relating to livelihood and promote the sustainable management of coppice woodlands and the promotion of their products. They are also a doorway to forums with conservation groups, wildlife trusts, local councils and planners.

- The Local Exchange Trading System (LETS) has the structure of barter to allow exchange of energy and time rather than money when it is unavailable or when one prefers to use it. It therefore fulfils the role of social capital and is an important buffer in times of hardship. The permaculture network forms a support system for people working towards, and helps to design, sustainable ways of living.

- The natural capital is an increasingly diverse resource base, providing me with many of my livelihood needs but also allowing a diversity of wildlife to prosper. Food, medicines, harvested rainwater, fuel, saps and building materials are all available from the land and surplus can be sold or bartered to bring in money or other needed items if my livelihood encounters difficult times.

- Human capital is also diverse showing a range of different skills adaptable to the different needs of the life of a forest dweller.

- The physical capital assets are formed by the structures and tools, etc. that have built up over time to ensure the viability of living the Woodland Way.

- Financial assets are primarily income from the produce of the land and creditworthiness, but can also be seen as investment into the land in the future, e.g. planting of fruit/walnuts for future markets (for more information on Prickly Nut Wood, see Example in Chapter Six).

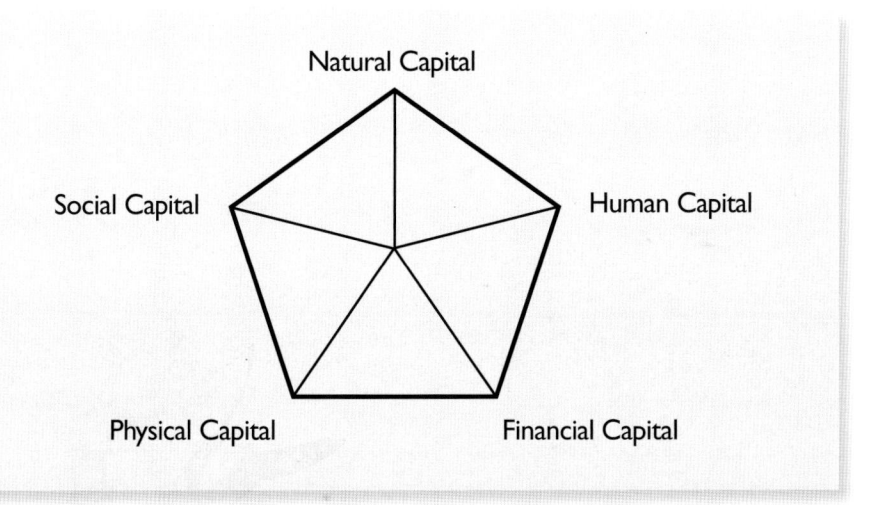

Left:
'Capital asset pentagon' from Sustainable Rural Livelihoods by Diana Carney.

26 The Woodland Way

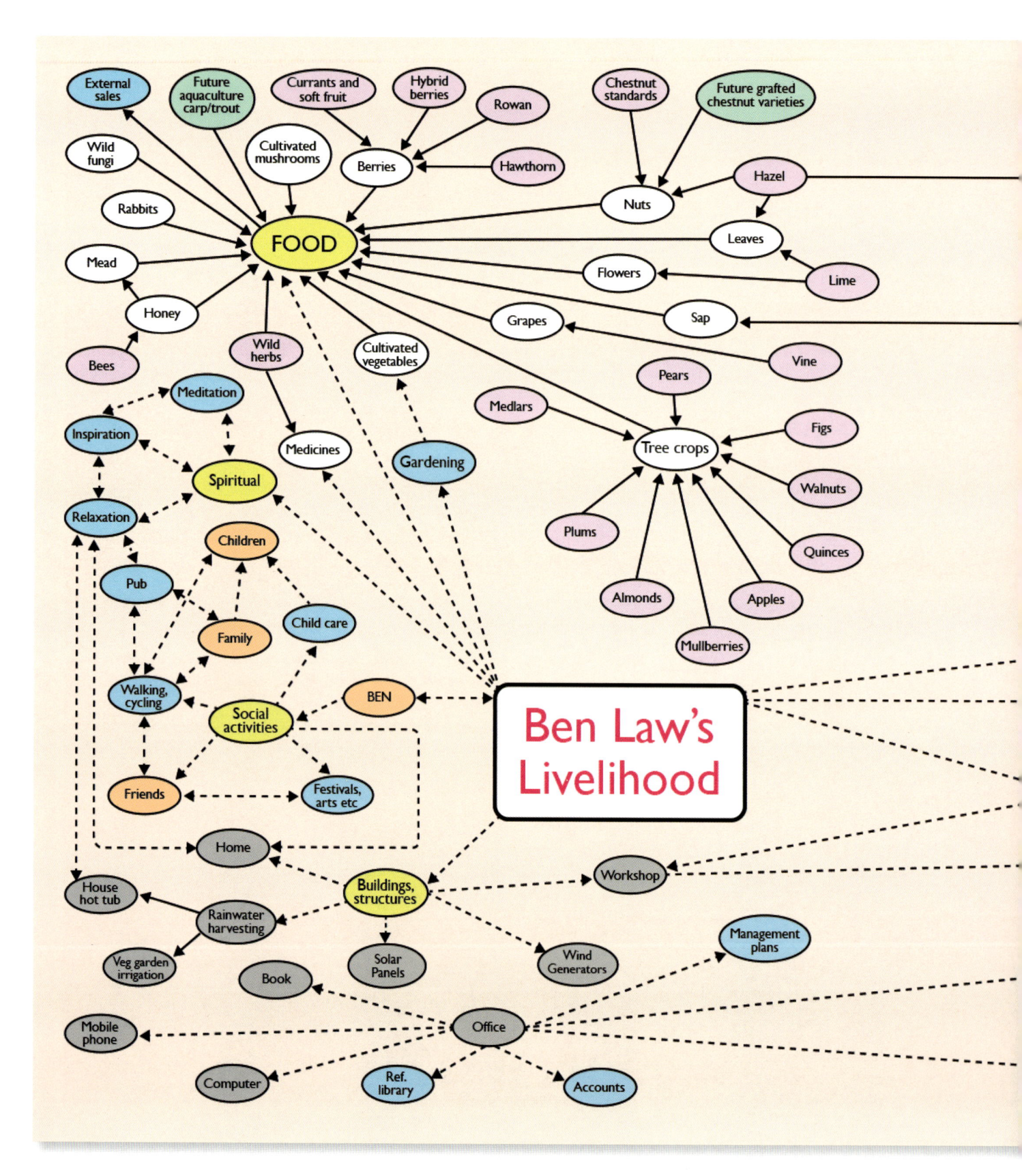

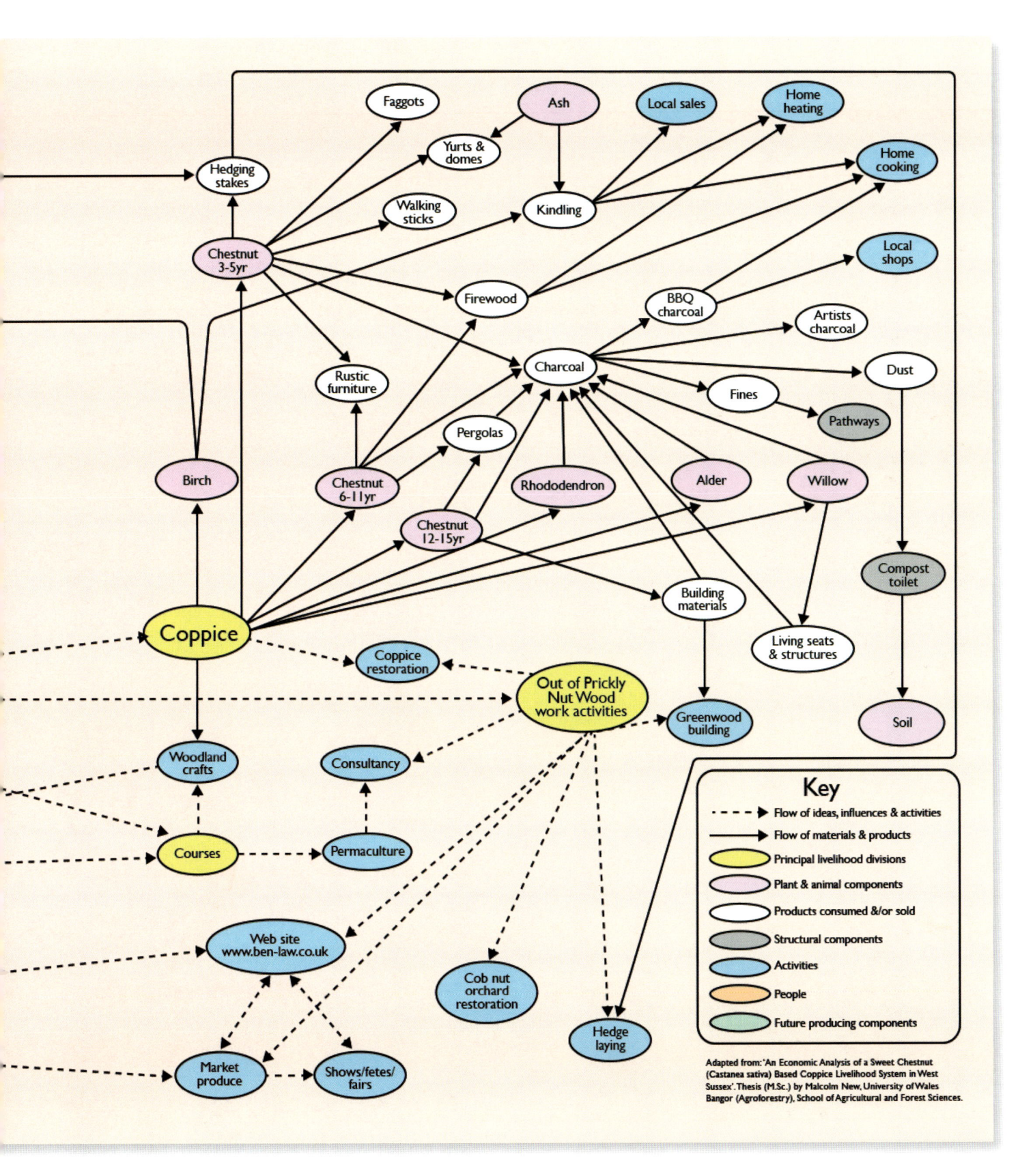

The 21st Century & The Return Of The Forest Dweller 27

PATTERNS OF FOREST DWELLERS

The pattern of forest dwellers is our clearest model of long-term sustainability from which to draw. In the table below I shall compare some common patterns of forest dwellers with those of the modern forester. The main similarity is that both groups earn their living from the forest.

The forest dweller's life is a holistic relationship with the woodland. The forester's relationship is conversely that of the predator, hunting the woodland for financial return, often exploiting the natural resource base of the woodland in the process.

THE RE-EMERGING WOODLAND COMMUNITY

"To dwell in a forest is to become part of its complex ecological functions. Eco-foresters then, are natives of the land." From *EcoForestry*, edited by Alan Drengson and Duncan Taylor.

Patterns of forest dwellers	Patterns of modern foresters
Live within the forest, usually part of a community	Live away from the forest, may or may not be integrated into a community
Build houses/structures from forest sourced materials, i.e. wood, clay, stone	Build or buy houses of materials often unconnected to the forest or local surroundings
Forestry is part of an integrated way of life	Forestry is work
The forest is home, and therefore looked after for future generations	Forests are entered to remove timber to make money; often clearfelled
Livelihood consists of utilising forest products; timber, wild food, medicines, saps, fibres and other forest products	Livelihood consists of travelling to forests, harvesting and marketing timber
Food growing and forest gardens in clearings	Food growing is not carried out within the forest
The forest is a map; every tree, stream, pathway is known by the inhabitants	Foresters often only know main extraction routes and areas of high value timber
Decisions based on years of observation, often generations of forest dwelling	Decisions often based upon one visit to the forest

Whether you are in favour of the prospect or not, more people are returning to live in the woods. This is not a new phenomenon, it is the natural return of a cyclical pattern within our society and it is up to us to make welcome those who are living in the countryside and earning a living from the woodland produce. As the chairman of a county council committee expressed to me, "At the end of the day, we have to support you. If the woods are not worked again and cannot be shown to be economic, we may lose them and the alternatives are the loss of our countryside and heritage."

In W.E. Hiley's *Woodland Management* (1967), which for many years formed the backbone of education for forestry students he discusses, "... coppice worked by woodsman who frequently lived in woods with their families, and as the youngsters learnt the craft from their parents, they carried on the family tradition and skill." It is to be expected therefore that as the amount of coppice being worked increases once more, the need for woodland families to reside in woodlands will grow. The price of housing in the countryside has inflated far beyond the relative increase in value of woodland produce and a large percentage of rural housing is occupied by people not engaged in working in the countryside. This does not need to be an issue. Freedom of choice is part of what makes our democracy, but freedom of choice needs also to be outstretched to the woodlanders as they return. It is therefore necessary to find low impact housing for those living seasonally or more permanently within woodland (see Planning Law and Woodlands in Chapter Eight). The obvious solution is wooden houses built from the woodland itself.

**Previous page:
Livelihood analysis of the author at Prickly Nut Wood.**

The process of reintegration needs to be two ways. The returning woodlanders need to be equally prepared to fit into and trade within the local community, have a respect and understanding of the life of local residents, and be able to live within the guidelines of low impact development.

As more people return to the woods, so the distances between people become increasingly smaller and the opportunity to share workload, orders and social activity increases. I await the day when a stroll down the footpath to the next coppice will reveal others with a kettle on the fire ready to share tales of the joy of the day that has passed.

There are also likely to be increases in woodland communities, like Tinkers Bubble in Stoke sub Hamdon in Somerset. Here a group of people with the interest of living sustainably and working the land together share the labour of the land and the patterns of daily life amongst themselves. Their commitment to sustainable woodland management includes: the non-use of fossil fuels; timber planked using a steam powered sawmill; and produce from their orchard which is sold at a local farmers' market. Provided the resources of the land can support the volume of people, it is hard to think of a more sustainable model for living than the woodland-based community interacting with their surrounding community through the sale of their produce.

Presently many woodlanders live in 'hiding' within the woods, doing their utmost to keep their way of life and position of their home a secret from the surrounding community for fear of being made homeless by the rigid structures of our present planning system. This leads to a growth in suspicion both from the woodlanders and from the local community and slows down the natural

Top:
A roundwood timber frame dwelling under construction in the woodland. Diddling, West Sussex.

Below:
Tinkers Bubble, Stoke sub Hamdon, Somerset.

The 21st Century & The Return Of The Forest Dweller

process of interaction and trading that should be taking place. Once awareness grows as to why people are living in the woods again and local communities begin to understand the history and culture of woodland life (and can put a face to the names of the woodlanders), people can take pride once more in the produce of their local woods and the relationship the woodland has with the surrounding village or town.

At the Stackpole Estate in South Pembrokeshire, owned by the National Trust, a residential site has been established within the woodland for a group of people to live and work the land. This is a positive and forward thinking approach by one of the country's largest landowners.

Low Impact Development

With the general need for more housing becoming higher on the agenda of every council, the pressure on rural areas grows. Each year sees another slither of green land removed as villages 'close their envelope' or stretch their boundaries to accommodate the housing pressure. Some areas are targeted for larger-scale development, changing the characteristics of the village and rural life within that area with one swift planning development. In my own county of West Sussex (as in many others), the government figures for new houses are far above what the planners feel they can accommodate. This strengthens the case for low impact development. These are developments that use the minimum of resources, have a low visual impact in the environment and don't carry with them the permanence of the bricks and mortar house. Rather than building another development of houses on the edge of a rural village, whose new occupants may have no associations with the village, surrounding

30 The Woodland Way

land or local rural economy, surely the case for low impact development, whose low impact occupants are directly working the land, or at least are contributing to the local rural economy, is a step far closer to sustainable development. In a situation where building a new brick house would not be allowed, the case for a low impact development made, for example, from locally sourced timber and tied to an agricultural or forestry agreement would open up an opportunity for many younger people who wish to work the land but who do not have access to it.

The rural planning group of 'The Land Is Ours', called Chapter 7 (taken from Agenda 21, Chapter 7: "To provide access to land for all households… through environmentally sound planning"), has been compiling a document aimed at creating some criteria for sustainable land use (see Planning for Sustainable Woodlands in Appendix Three). Presently land-based projects such as forestry and agriculture are assessed on their economic viability (money earned). Although this is desirable, it is far removed from a definition of sustainability as the DFID capital asset pentagram clearly shows. The draft criteria in Appendix Three offers guidelines for sustainable land-use. These are guidelines that should help planners to differentiate between those who genuinely wish to work the land sustainably and those whose only interest in planning in the countryside is land speculation.

A Case For The Cabin?

Low impact woodland developments should include benders or yurts. Both are easy to construct and to move in a woodland and have far less of an impact on the woodland than a caravan, which has to be towed into the woodland. Caravans, of course, have their place and they are well embedded within the framework of planning law and are presently the most likely of temporary structures to be accepted by a council on a planning application.

Caravans constructed from local timber are far more aesthetically pleasing than aluminium touring caravans and blend into the woodland environment. Wood structures, such as cabins and timber framed wood buildings, are the obvious choice for longer-term low impact development. What could be more unobtrusive in a woodland than a building constructed from the same material as the growing trees? Cabins are linked with woodlands and woodland workers across the globe. The reason for their popularity is common sense. Many cabins are built deep within woodlands where other materials cannot be transported, and if the building materials are growing around where you need to reside surely it makes the most logical and low impact sense to use them!

We find ourselves at an interesting point in planning history. The need for more housing development is being pushed by central government, whereas individual counties are trying to resist development in rural areas. Forestry land is predominantly either commercial plantations or neglected woodland (much of which is coppice). A strong resurgence in traditional forestry practices calls out for more acceptance of woodland dwelling and an understanding that woodland dwelling is part of our cultural heritage. Local Agenda 21 calls for the sustainable management of our resources, while the modern forestry model is based on monetary value and short-term thinking and not on the sustainable management of our resources. The acceptability of low impact forest dwelling linked to traditional management and modern marketing seems the most sensible way forward.

Above left:
A roundwood cabin, ideal woodland accommodation, West Sussex.

Below left:
A Prickly Nut Wood roundwood caravan, ideal temporary or seasonal accommodation.

Above:
Woodland cabin, Leckmelm Wood, Ullapool, Scotland.

WOODLAND ASSESSMENT & MANAGEMENT PLANNING

Chapter Three

I am often greeted with the words, "We've got this woodland and we are not quite sure what to do with it. Should we leave it, or does it need managing?". The answer may be to leave it or may be to manage it, but the process of getting to that decision will be the same.

Old Records

Before visiting a woodland, thorough research into the history of the woodland will reveal much about its past. Old people who have lived in the area for most of their lives may at least shed some light on the past 60 years or so as to how it has been managed. Large estates often have detailed woodland records going back 100 or 200 years, and the name of the woodland itself can give away much. When looking at an Ordnance Survey map many woodlands still contain the word 'common' even if commoners' rights are no longer registered. The is a clue that it might have been a wood pasture at some point in history. The word copse refers to coppice and even if the wood stands now as a coniferous plantation, its past history remains in its name.

Documentation of old woodlands is sometimes available from the County Records office and occasionally from reference libraries. The scripts and information are not always easy to decipher but useful items like local estate maps may be helpful in sourcing the woodland's origins. The earliest Ordnance Survey maps date back to the beginning of the 19th century, and these clearly show the position of woodlands at that time. Also of use are maps of Artefacts (usually referred to as bank and ditch maps) which give the position of man-made earthworks. Identifying earthworks and deciphering the maps can be made easier with the help of a field archaeologist, but many ditches and banks in woodlands are clearly visible to the less trained eye. Aerial photographs can be useful for showing the layout of woodland rides, glades, streams etc.

Winter photographs are more useful for broadleafs and, with coniferous plantations, photographs may be of use to identify species change. Identification of the woodland flora will be of help in understanding the patterns of evolution of the woodland by species. A record of plant species will also be a useful indicator of increase or decrease in biodiversity when a future survey is carried out. Plant species also give indicators of what other species may be available, i.e. violets in a coppice woodland show a suitable habitat, for example, for pearl-bordered fritillaries which are one of the fritillaries which feed on violets during the larval stage. Flora surveys should be carried out at many different times of the year as certain plants, like wild garlic, only appear visibly for a short period of time.

Observation & Recording

When visiting a woodland give yourself plenty of time. Allow how long you think you are going to need and then double it! Take a map and a notebook and make some records of what you see. Has there been any recent forestry activity? Are there any well trodden paths? If so, are they human or animal? Look for tracks (easiest in snow), droppings, nests, holes in trees and start to build up a record of the woods. Record tree and plant species, look at the age of trees, have they been coppiced? Is one species dominant? Is the planting regular? Look for ditches, banks, earthworks, excavations and unusual trees. Stop and sit in one place for a while, get a feeling for the woodland and see what comes to you.

Don't just make one visit; each visit to a woodland increases your understanding and gives you a clearer picture of how it has evolved and what to do next.

Main habitat preferences of butterflies breeding in British woodlands

Very open sunny rides or glades (less than 20% shade)

Chequered skipper
Small skipper
Essex skipper
Large skipper
Dingy skipper
Grizzled skipper*
Brimstone+
Orange-tip
Green hairstreak+
Brown hairstreak+
Small copper
Brown argus
Common blue
Holly blue+
Duke of Burgundy*
Painted lady
Red admiral
Small tortoiseshell
Peacock
Comma
Marsh fritillary
Small pearl-bordered fritillary*
Dark green fritillary*
Wall
Scotch argus
Marbled white
Hedge brown
Meadow brown
Small heath

Lightly shaded rides or glades (10-40% shade)

Ringlet
Wood white

Fairly heavily shaded rides or glades (40-90% shade)

Speckled wood
Green-veined white

Newly cut woodland (including ride margins and coppice)

Pearl-bordered fritillary
High brown fritillary
Heath fritillary

Dappled shade within woodland or along edges of wood or rides

White admiral
Silver-washed fritillary

Tree or shrub feeders, mostly confined to the canopy

Purple hairstreak
White-letter hairstreak
Black hairstreak+
Purple emperor
Large tortoiseshell

* these species also breed in newly cut woodland
\+ these species require particular trees/shrubs.

From *Woodland Rides & Glades*, reprinted with kind permission of the Joint Nature Conservation Committee.

Opposite left:
Old earthbank in ancient coppiced woodland, Lickfold, West Sussex.

Visit at different seasons and add to your plant and tree species list and make sure you visit at night. The wood comes alive at night and you will learn much through sounds and sights in the moonlight as to who else inhabits the woodland.

Study the access to the woodlands, as good access is the key to viability if timber produce is to be extracted from the woods. Look at the topography and soil type as this will also affect the access at different times of year. If you don't know the average local rainfall, contact your local meteorological office and they can give you at least the last 10 years' monthly rainfall figures to work from. The results from your visits and the information you have collected should give you a woodland survey which should cover:

- The name of the woodland
- Area of the woodland
- Map reference of the entrance to the woodland
- Ownership of the woodland
- District Council
- Parish Council
- Local Forestry Commission Office
- Any legal conservation status (SSSI for example)
- Species lists for trees, other plants, birds, mammals, butterflies, insects etc.
- Altitude
- Soil type (test soil in different parts of the woodland/check soil maps)
- Aspect
- Degree of slope (important for extraction routes and identifying frost pockets)
- Water courses

- Drainage
- Unusual features
- Climate
- Microclimate (trees can create a variety of different microclimates some of which can be prime growing areas for sun loving species)
- Annual rainfall
- Access for vehicles
- Access by foot or horse
- Public footpaths and bridleways
- Woodland rides – size and condition. (Is there a solid base for mechanised extraction or is horse extraction a better option? Is there a diversity of species growing on the ride which may be important butterfly food or indicators of other rare species?)
- Any recent forestry activity
- Past patterns of management
- Potential problems – invasive species – deer, rabbits, squirrels

Also with this survey you should make your own 'resource reality' checklist which should include:

- What is your long-term vision for the woods?
- Your own skills and resources (including financial)
- Amount of time available
- Local markets for sale of produce
- Your knowledge
- Your objectives for the woodland
- Other people's skills that may become involved
- Available grants (see Chapter Eight)
- Local and national organisations
- If at any stage you feel at all uncertain about how to proceed, do seek professional advice.

Assessing Woodland Flora

Assessing the history and age of a woodland by studying the flora is a skilled and area-specific methodology. However, the woodsman can build up a useful knowledge of indicator species, the most useful being species that can survive in the shade in the heart of the woodland and have limited colonising ability. I am not going to offer a comprehensive list of ancient woodland indicator species, as the specific location of a woodland means many variations of indicator plants occur. My advice is to read up on your local natural history and visit known local ancient woodlands and build up a picture of the type of plants and plant combinations, you may encounter. The more of these indicator species present within the woodlands local to me indicates a higher probability of an ancient woodland.

Access

If timber is to be extracted from the woodland, the condition, distance and slope of the rides will have a major effect on the economic viability of the enterprise.

Above left:
Deer prints highlighted by the snow.

Above:
Honeysuckle clearly seen in January is an important food plant for the White Admiral butterfly.

Some indicator plants I look for in woodlands in my locality

Anemone	*Anemone nemorosa*
Bluebell	*Hyacinthus non-scripta*
Butcher's broom	*Ruscus aculeatus*
Dog's mercury	*Mercuralis perennis*
Early dog violet	*Viola reichenbachiana*
Early purple orchid	*Orchis mascula*
Hellebore	*Epipactis helleborine*
St. John's wort	*Hypericum hirsutum*
Sessile oak	*Quercus petraea*
Solomon's seal	*Polygonatum multiflorum*
Wild garlic	*Allium ursinum*
Wild service tree	*Sorbus torminalis*
Woodruff	*Asperula odorata*
Wood sorrel	*Oxalis acetosella*
Wood spurge	*Euphorbia amygdaloides*
Yellow archangel	*Galeobdolon luteum*

divides such as streams, earth banks or rides or a clear change in vegetation or topography. Use these natural divides where they exist. You may need to create new extraction rides and these will form new divides to help create different sections. Take each section and work on a plan for 10 years, always taking into account its relationship to the other sections and the long-term vision as a whole.

If felling is to take place, it is most likely you will need a felling licence (see Chapter Eight). This can be obtained through the Forestry Commission for a particular section or the whole woodland could become part of a five year plan in a Woodland Grant Scheme application, which will include all felling licences. By this time, you will have done all the work necessary to fill out the application forms and a visit from a Forest Commission Officer should reassure you of your long-term objectives.

Most woodlands that have been managed for timber or coppice have an established network of rides, consisting usually of a main ride which has a hard base and a number of narrower branching rides. Sloping ground rides should, where possible, follow the contours of the land. The time and energy used to extract timber from awkward situations should not be underestimated. Also the cost of establishing a new network of rides may take many generations of woodland work to repay the investment.

If large timber for sawlogs is to be extracted and the condition of the rides is poor, then mobile sawmilling and/or horse extraction should be considered. If the woodland is overstood coppice, then on-site conversion to charcoal should be considered as an option, as the weight and volume of the timber will be greatly reduced. In woodlands with difficult access, once the coppice is back on cycle, it can then be kept to a short rotation so that produce is kept small and light. If improving access rides, try to use local materials such as 'sandstone hoggin' or 'limestone scalpings', depending your local stone.

The next step is to analyse your survey and work out your objectives for the woodland in the long term; and proposals or work to be undertaken (if any at all) to meet these objectives and a timescale to work within. Start with your long-term vision for the woodland: how you imagine the woodland will look in 100 years' time, and then consider how the woodland will be in 10 years' time. A plan can be formulated year by year to achieve the results over 10 years, always keeping the longer vision in mind. If the woodland is large, it may help to break it down into sections. The woodland may have natural

A First Assessment

I was once asked to assess a local woodland by a new owner. The owner asked me because I was local and had some existing knowledge of the woodland and its past history. The woodland forms part of the same Site of Special Scientific Interest as Prickly Nut Wood. Therefore it is likely there will be an overlap in management objectives. I know many of the woodlands in my locality, especially those which have footpaths passing through.

The woodland is primarily coppice, cants of sweet chestnut and also some ancient oak standards over hazel understorey and two plantations of approximately 25 years of age containing mainly European larch and Scots pine. I know this information already because I am familiar with the local woodlands.

38   The Woodland Way

Notes From The Walk

Walking along the main access ride of the woodland, I feel a sense of neglect compared with the nearby coppice in which I work.

Having an interest in local woodlands, I have obtained an old map which gives old names (as opposed to the more commonly used numbers) for the different cants within the woodland.

The first cant I encounter is Stone Pit Piece, named from a very small sandstone quarry located above the cant. The cant is of sweet chestnut with an influx of rhododendron. The cant was last cut about seven years ago and the coppice regrowth is good, forming a well stocked cant with many good poles for potential markets. The cant rises steeply to the west and the quality of coppice declines. Felling of the coppice to the far west could be awkward, possibly hazardous, and I would consider singling the best stems and allowing them to grow on for

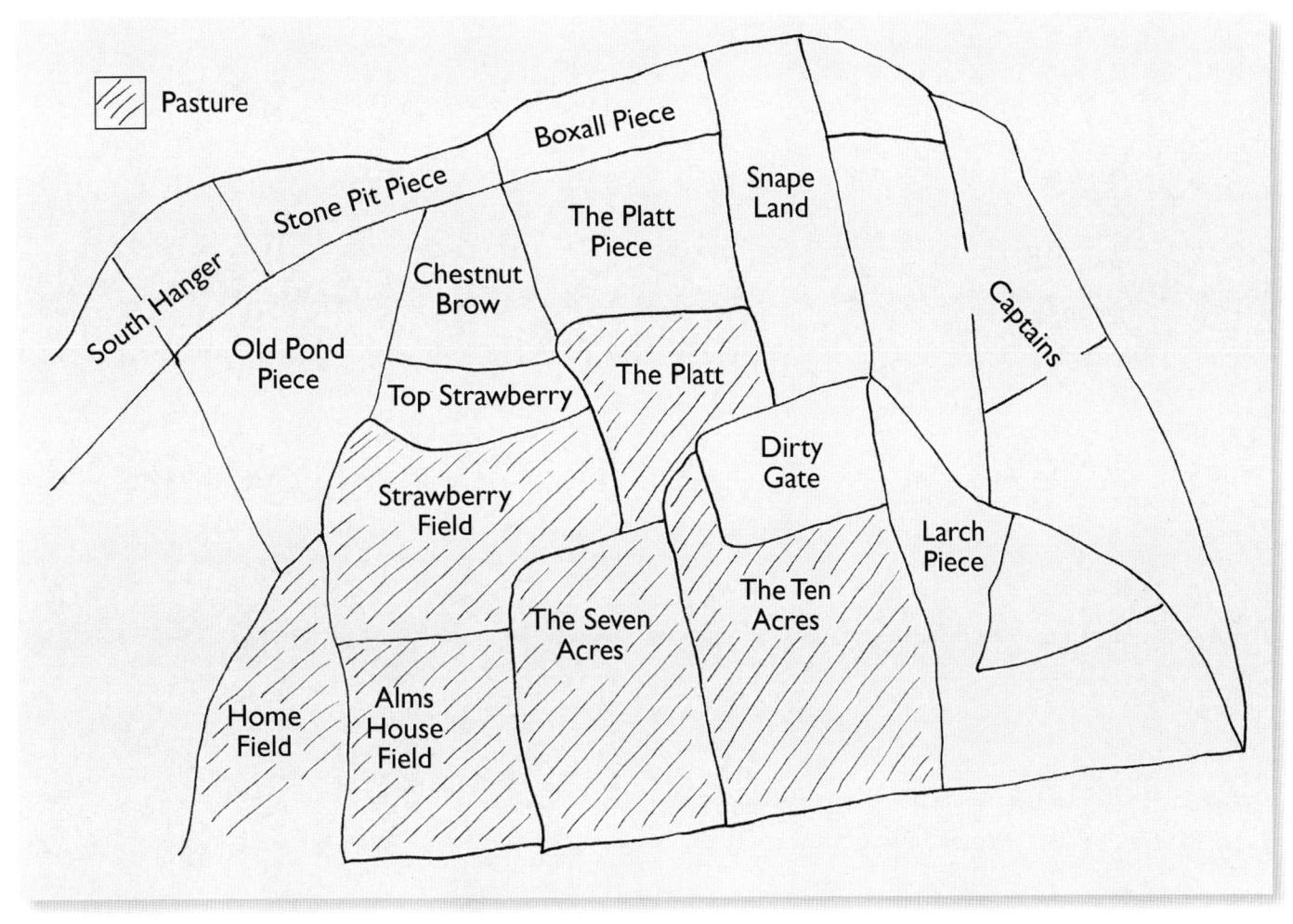

Opposite page
Top left:
Butcher's broom (showing its rarely noticed flower).

Top right:
Wild garlic.

Centre left:
Common dog violet.

Bottom left:
Early purple orchid.

Bottom centre:
Wood spurge.

Bottom right:
Wild service tree showing its Autumn glory.

This page
Above: **Named cants in the woodland, based on an old map.**

nut production. This would leave 80 percent of the cant on coppice production and allow the 20 percent with the steepest slope and furthest distance from the ride to be used for the October nut harvest.

I continue along the main ride and arrive at Boxhall Piece (Boxhall is a well known family name in my area). The cant contains very overstood sweet chestnut with a few oak standards. The stools of the overstood sweet chestnut are spaced far apart and many have been windblown from not being coppiced. There is evidence of past management. A few stems have been felled and cleaved, presumably for fencing rails. The cleaved wood shows that the grain in the wood has twisted and therefore the wood has been left. There is a badger set, well concealed in one corner and clearly active with fresh diggings, claw marks and well used pathways. The viability of recoppicing this cant and creating a productive, on-cycle cant is dubious due to the wide spacing and the amount of planting and layering that would be needed. There is also the badger set to consider and an intensive recoppicing and replanting programme may well disturb the occupants. A better option would be to single the best chestnut stems and grow them on for nut production. The felled timber could be inoculated with mushroom spawn and hence convert a piece of derelict coppice into a food production woodland. This would cause less disturbance for the badgers and would be a minimum intervention approach producing a long-term nut crop.

The next cant is the Platt Piece. At a first glance this seems quite a mixture, and it takes me time and observation to sense what has occurred in the past. I sit under a mature larch of about 60 years of age and breathe in the clear woodland air. I remember a strip of chestnut about three stools deep being cut adjacent to the ride about four years ago for walking sticks. Beyond these stools there is a small area of about eight year-old sweet

Left:
Derelict chestnut coppice in Boxhall Piece.

Below:
Hazel's distinctive yellow catkins.

chestnut and beyond this I can see the distinctive yellow tails of hazel catkins.

As I venture deeper into the cant I find myself amidst overstood ash and hazel coppice. The stools look healthy with many living stems per stool. With some layering, the stocking distance could ensure good yields of straight poles for the future. The access is poor and to convert the overstood coppice to charcoal within the cant seems the most viable option at this stage, but more visits and observation may reveal other possibilities. There is also about an acre of fair condition, 16 year-old sweet chestnut ready for cutting for fencing or building materials. The bracken looks well established to the south of this chestnut and has colonised a small clearing.

The ride descends towards Snapelands. This is an area of mainly sweet chestnut coppice of varied ages and well established rhododendron. Snapelands is an old word for 'boggy ground' and the erosion on the ride adjacent to this cant clearly shows the amount of water that runs off from Bexley Hill above. Many of the chestnut stools show extensive squirrel and rabbit damage which will affect the quality of produce made from the poles. There are stacks of pointed chestnut fencing stakes close to the ride. This could be the sadly familiar tale of a coppice worker/landowner dispute or a lack of available markets.

Entering Captains is a refreshing contrast from the cants of sweet chestnut. Oak standards dominate the landscape with a rich understorey of hazel, ash and field maple. Many of the standards are large spreading trees and have shaded out a lot of the underwood below.

The hazel is struggling to survive in many places with only a few stems per stool still alive. Recoppicing will need to be carried out as an urgent priority if the hazel is to survive. This would need to be accompanied by the reduction in the number of standards to allow more light to the understorey. 'Stop planning and observe', I have to remind myself.

I pass butcher's broom, an ancient woodland indicator. I will return in late April and record the spring ground flora.

Larch Piece lives up to its name. The larch was planted 25 years ago and has never been thinned. There is evidence of windblown trees and a thinning at the earliest opportunity would give the better stems a chance to grow on. Some ground flora is evident where the fallen trees have allowed the light in, but under the density of the plantation, there is little evidence of life.

At Dirty Gate, I encounter a stream and alongside it an abundance of alder coppice. I estimate it was last coppiced about 30 years ago. The stems are good and straight and could be planked for flooring. I sit down by the stream and listen to the bubbling flow of water. Access is poor, so extraction is unlikely to be viable, even bringing in a mobile mill and taking out planks could be awkward. I consider the charcoal option. Alder makes a good quality charcoal and taking out the reduced weight product would be viable. Mushrooms are another option. The logs could be inoculated with mushroom spawn *in situ* and the stream could be used to shock them into fruiting by immersing them in it (see Fungi in Chapter Seven). This might be the best solution.

As I return along the main ride, my main feelings are of how little I know the wood and how much more time and observation is necessary before I can draw up a plan.

This is how I begin a woodland assessment. I already have some previous knowledge of the woodland and have walked through it on a few occasions. On this walk I ventured deeper, but I am still only beginning to learn about the woodland, its past and its species. There are many lessons to be learnt from the past which shape our plans for the future and woodland assessment is not any different. I must return for many more visits and search out past workers and local contacts who will have more knowledge on the recent history.

Assessment is a skilled and enjoyable activity. Building up a picture of a woodland, getting to understand it intimately takes time, realistically a lifetime. Remember to be flexible as what seems appropriate at the time of writing a management plan may change as you begin to work in, or visit, the woodland more regularly. Ask other opinions, as different eyes and knowledge can open up a wider understanding of the woodland. Above all, take time to observe. A woodland may have taken hundreds of years to reach its present situation, only to be changed on the whim of a quick visit.

Opposite left:
Autumn colour of European larch.

Right:
Catkins of the common alder.

ESTABLISHING NEW WOODLANDS

Chapter Four

Natural Regeneration

The easiest way to establish a new woodland is to allow the soil and climate to make the species choices and establish the woodland for you through natural regeneration. The natural process of succession from arable or grassland, through pioneer scrub species to pioneer trees to climax woodland is the way nature establishes new woodlands. All you need to do is fence out rabbits, deer and livestock and check there are some mature trees nearby for seed stock and let nature do the rest, and we can also get a grant to encourage us to do this! (See Chapter Eight.)

Observing the process of natural regeneration is in itself a unique and fulfilling experience as what we are observing is nature reclaiming our cultivated low cropped arable or grassland and allowing the land to reach its natural climax. It is a process of watching a low cropped grassland grow long and tufty and the first pioneers of often bramble, gorse and bracken start the process of vertical growth, while hedgerow trees begin to extend outwards colonising the newly found space. Birch, alder and willow are often some of the earliest pioneer trees and will create the first canopy before the climax species eventually burst through. The type of species is dependent on all types of climatic factors and what is growing nearby. When the climax canopy is formed, nature develops no further and continues the cyclical pattern. A dead tree falls, allowing light through the canopy to the forest floor enabling young waiting trees to extend upwards towards the light. And so the process repeats itself.

Our options with a naturally generating woodland are to allow the full process to climax woodland to occur or to intervene and select particular species and adjust stocking density during the evolution process. We can also interplant into a naturally regenerating woodland utilising pioneers such as gorse to fix nitrogen and create shelter for our chosen species. Natural regeneration gives us a lot of options and shows us what really wants to grow on the land. It is also the option of minimum input for maximum gain.

Planting

Before planting a new woodland it is important to take time to be clear and certain that you have thought through the long-term results of the planting. Planting woodlands has a very dramatic effect on the landscape and it is many years before these results are clearly realised.

Be clear of the purpose for planting a wood. Is it aimed at a particular market in the future? Is there a diversity of species in case the market has lapsed in, say, 80 years time? Are the chosen species suitable for the soil type and for the local climate? Is planting trees or natural regeneration the best option?

There are occasions when planting rather than allowing woodlands to naturally regenerate is the chosen option. I have planted a few woodlands specifically with a long-term plan to create a multifunctional woodland, which will find new markets for its products as well as supplying food. Traditional woodlands have always been places to forage. My local common supplies me with crab apples, sloes, rosehips and blackberries. When I design woodlands a food element is always an important function. These are, after all, new woodlands where nature will replicate the old and no doubt adapt whatever I plant, but in a time when our wilder fruit trees are becoming scarce, I am keen to start with the emphasis not solely on timber.

Designing Woodlands

For me, one of the most inspiring transitions in my understanding of woodlands came through attending a permaculture design course. Permaculture brought together all the separated disciplines of land use, care of the environment and social justice, and connected them into a coherent framework: a design process. My woodland life has been an exploration of this process and Prickly Nut Wood my home and my college of learning.

Permaculture contains principles of design evolved by Bill Mollison and David Holmgren that have been tried and tested over a number of decades to produce a design system for sustainability. The principles can be used to design agricultural systems, gardens, houses, communities, businesses and, of course, woodlands. Any well designed permaculture system will be ecologically sound and economically viable and will minimise inputs and recycle all potential wastes within the system. The design is applied common sense.

The following section looks at permaculture principles in relation to woodlands (for more information on permaculture, see the Bibliography in Appendix Seven).

Opposite left:
Birch natural regeneration is encouraged by the author amongst and between chestnut cants as it adds diversity.

Permaculture Principles as Applied to Designing Woodlands

Observe & Interact

Observation and interaction is a cyclical pattern of improving the design of our landscapes. With woodlands, and in particular coppice woodlands, we have a cyclical pattern of management and observation carried out over generations of woodsmen. If we take a woodland like Bradfield Woods (see page 7) which we know has been coppiced for over a

Below:
A swale dug by the author for Meera's Wood in Lodsworth, West Sussex prior to planting a new woodland.

thousand years, throughout that time different woodsman have observed and coppiced and improved their management until we arrive at the point today where all those generations of practice have created a rich bio-diverse habitat. The observation cycle continues and new woodsmen will take on the management with the knowledge and decisions of past woodsmen all woven into the fabric of the woodscape.

Relative Location
'Put the woodland in the right place'. What is the location of the woodland in relation to houses, water, wind, sun etc.?

Is the planned position of the woodland benefiting other enterprises in the surrounding landscape? For example, is its planned position to create a shelterbelt for a microclimatic area for growing other crops? Or will it cause shade to buildings or crops etc.?

Each Element Performs Many Functions
What are the different elements within the woodland and what functions are they performing? For example, a pond within the woodland could be performing the functions of fire control, habitat creation, reservoir for irrigation of young trees, growing of aquatic crops, and a place of contemplation. The woodland as a whole could be performing functions of recreation, timber production, food production, wildlife value, shelterbelt, wildlife corridor and carbon fixing.

Each Important Function Is Supported by Many Elements
To continue my previous example, the important function of irrigating young trees should be supported by more elements than the woodland pond. Swales (contour ditches to catch rainwater and increase water infiltration to the soil) dug within the woodland and rainwater harvesting off any forestry buildings would support the function of irrigation within the woodland area.

Energy Efficient Planning
Efficient energy planning includes placing elements according to how much we need to use them or how often we need to visit them. For example, in a new woodland planting, it would be more energy efficient to place food producing trees on the edge of rides and regular walkways where they can be harvested easily and will catch more sun to ripen fruit. Trees which are being grown for long-term timber value and will need little maintenance, can be placed further away and more in the middle of plantings so that they will be drawn up by other species.

It is important to consider incoming energies such as wind and to divert it away from areas where we do not want it, such as glass houses and tender crops, and direct it towards areas where we do want it such as wind turbines. It is important to consider the site in profile and look for degrees of slope that can be used to gravity feed water or signify the direction of flow of frost. Slope and the use of contours will help establish rides and suitable non-eroding routes for future timber extraction.

Using Biological Resources
Wherever possible, biological resources are used to save energy. In a woodland, the model of natural regeneration where trees plant themselves is a preferable energy-saving choice over human planting. Other biological resources which can be used are pigs to prepare the land by their natural 'digging' actions and at the same time they will fertilise the land with their manure in preparation for planting (see Wild Boar in Chapter Seven). Leguminous plants such as clover

can be broadcast prior to planting to help fix nitrogen, and the planting of nitrogen fixing trees such as alder will continue to fix nitrogen throughout the establishing woodland.

Energy Cycling, or Catch & Store Energy

Most woodlands in Britain involve growing timber trees for a future market and often the timber travels great distances to sawmills and then retail outlets. The trees themselves have often come from seedstock obtained from European or American sources, fertilisers are added and herbicides used, often bought in from a company manufacturing chemicals in some distant part of the globe. All of this, when translated into the figures of short-term economic thinking, looks good for the woodland owner who can see a large financial return when the trees are felled in perhaps 50 or 60 years time. However, the hidden costs to the environment have not been included in the economics. The amount of pollution caused by the transport distances of all the components that have been utilised to create the finished product is not accounted for.

In permaculture design we look to eliminate these energy inefficient transport distances, stop the flow of energy leaving the woodland environment and turn it into cyclical energy. Hence a permaculturaly designed woodland will aim to collect locally provenanced seed (or use natural regeneration if possible) to establish the woodland. Any fertiliser input will come from biological resources, such as from planting green manures or using animal manure from as near to the woodland as possible. The leaves from the trees can be used as fodder to feed the livestock that produce the manure, using techniques such as shredding.

Any thinnings can be used for firewood or fencing and, in turn, the finished timber can have on site added value and be marketed locally to ensure the minimum energy loss and as much energy cycling as possible. Timber produce can be bartered locally in return for labour in the woodland which completes the energy cycle.

With wild energies, such as wind and water, entering the woodland, permaculture design works to catch, store and then utilise these energies as many times as possible before they leave the woodland. Hence a series of ponds can be created to hold water on a hillside to be gravity fed to crops below when needed. Wind can be converted into electricity via a wind turbine for future use, as can the sun through solar panels, perhaps in a woodland workshop to make added value products or to power a light and computer such as the one I have used to write this book!

Another store of energy at Prickly Nut Wood is 'laying down' timber for the future. As I restore the derelict coppice, I select standards and mill up those with a good figure into planks. These planks are layed down in my drying shed to air dry for 12-25 years. Air dried timber fetches a premium over kiln dried and

Opposite left:
Water from the workshop at Prickly Nut Wood feeds a 10,000 litre underground storage tank, which in turn irrigates the vegetable garden.

Below:
Solar panels on the workshop at Prickly Nut Wood.

when I am too old to carry out the physical graft of working in the woods itself, I plan to retire into my workshop and turn some of these planks into fine tables and furniture. The planks I am laying down today will become my woodland pension.

Small-scale Intensive Systems
Permaculture looks towards human-scale systems. Last summer while enjoying the pleasures of one of England's beautiful network of footpaths, I came upon a modern harvest scene. A bailer had broken down and the tractor driver was in an intense telephone discussion with, I presumed, an agricultural engineer. The discussion seemed to be about the hydraulic system that was no longer working and when it could be fixed. The sky was beginning to darken and I could smell the scent of rain on the air. The driver was also aware of this and hence his tone of voice and rational patterns of thinking deserted him when the engineer suggested a repair the following day! The tractor driver was helpless without his machine because the bales he had produced were the large round variety, too large to be handled by human labour. With traditional sized bales, I could have lent a hand and we would have had them stacked and covered before the skies fully opened!

The large forest harvesters and huge transport trucks of the forest industry have no place in a permaculture system. Permaculture systems are designed to use hand tools and small-scale fuel tools. A patchwork of small-scale diverse woodlands, linked together to form the larger forest serving their local villages and towns with the timber and produce they need, and community woodlands linked around and within the cities, should be the design for a sustainable future.

There is a technique we can use within our establishing woodlands, called 'time stacking'. From a greenfield or arable situation to a closed canopy woodland may take between 10-20 years depending on species, planting distances, soil condition etc. So there is a window of time during which the woodland is establishing that we can utilise the land to grow other crops; annuals, perennials, shrubs and introduce livestock, such as poultry, as the system establishes. Once the canopy is closed there will be a limit to which other crops can be raised, but during the 20 year period, many years of produce will have been harvested and the soil will have improved through poultry and green manures.

Many of our needs can be gained from a small-scale intensive woodland. The woodland can provide our timber needs for buildings, fencing and firewood, and some of our food needs can be provided from the fruit, nuts and saps. It can provide some of our medicines, and create an environment in which to relax and contemplate. If we can gain our needs from small-scale woodlands, it will free up more land for other species and for nature's evolution. This is land where nature designs herself. Such areas which we loosely refer to as wilderness are an essential part of any permaculture system. Wilderness areas are where we are the visitors rather than managers and it is such rare environments that remain as indicators, a benchmark of comparison as regards species diversity and patterns of evolution compared to our managed landscape.

Apply Self Regulation & Accept Feedback
Working a woodland is a continual lesson in monitoring what has been successful and what has not. If I thin the larch plantation unevenly, I will create opportunity for wind throw, and the woodland will clearly point this out. Prickly Nut Wood is a Site of Special Scientific Interest for its Bryothyte community (mosses, lichens and ferns). Every ten years a survey is carried out to determine what species are there. If the results of a new survey showed a serious decline in a species compared to a previous one, I would need to accept that feedback and review my management techniques.

Accelerating Succession & Evolution
Succession is how a woodland is established through the process of natural regeneration. If a garden lawn that has been carefully mown year after year, is left and not cut, it will grow long and rough. Over time a herbaceous layer will establish and brambles are likely to appear and then a shrub layer, followed by pioneer tree species such as birch. Eventually it will evolve into a woodland.

With an understanding of the pattern of succession, we can look at introducing useful species during the process of natural regeneration. We can add legumes to help fix nitrogen and plant comfrey amongst the herb layer to create a yielding crop for making biological fertiliser.

We can add useful tree species, such as walnuts, that are unlikely to appear through natural regeneration to accelerate the climax layer.

Diversity
I have been fortunate enough to spend a short period of my life in the Amazon rainforest. I found myself within an ecosystem of a vast and rich diversity of flora and fauna – and that is only what I could identify through the human eyes of a noisy intruder! The hidden insect species, let alone the vast diversity of activity

going on below the ground, I could only imagine. The yield of this forest, and its great river system containing 20 percent of the world's fresh water, is not measurable in human terms. We are only just beginning to realise its vast resource in the form of medicinal and non-timber crops let alone its carbon fixing properties. Despite our impressive research and scientific plant classifications, it is from the diverse culture of forest dwellers that we are learning the uses of plants. We are also finding examples, that where we have classified a particular type of tree, indigenous people's local knowledge shows that there are three trees. The differences are so subtle we have not been aware of them and yet the forest dwellers have a variety of different uses for the different trees we thought were the same!

Walking through such a diverse forest might bring feelings of 'untidiness and disorderliness' to a commercial forester used to the overtidy and orderly plantations of the British Isles. But tidiness is not a characteristic of nature; in fact nature thrives on disturbing our attempts to control her. Diversity in life, whether in plants, activities or culture is a principle of permaculture design. The yield of a monocultural plantation is timber. A diverse woodland will yield timber, nuts, fruit, medicines, saps, oils, resins, fibres, habitats, leaf proteins, animal proteins, as well as a place to relax and contemplate that which is greater than ourselves.

The yield of timber from such a diverse woodland may be less than from a plantation but the sum of the yields of all the produce will be greater.

Diversity of species also ensures a far lower risk of disease. Within a diverse woodland a tree may die but that tree will be utilised by many other species and is part of the whole intricate web of the woodland. In a plantation, a species-specific disease can wipe out the whole forest as all the trees are often the same species. Diversity below the ground should also not be forgotten. An orchard with perhaps thirty different apple varieties may only be grafted onto one root stock, hence a monoculture exists below the ground and disease risk is increased.

Above:
A lobular pattern design cutting through an existing landscape.

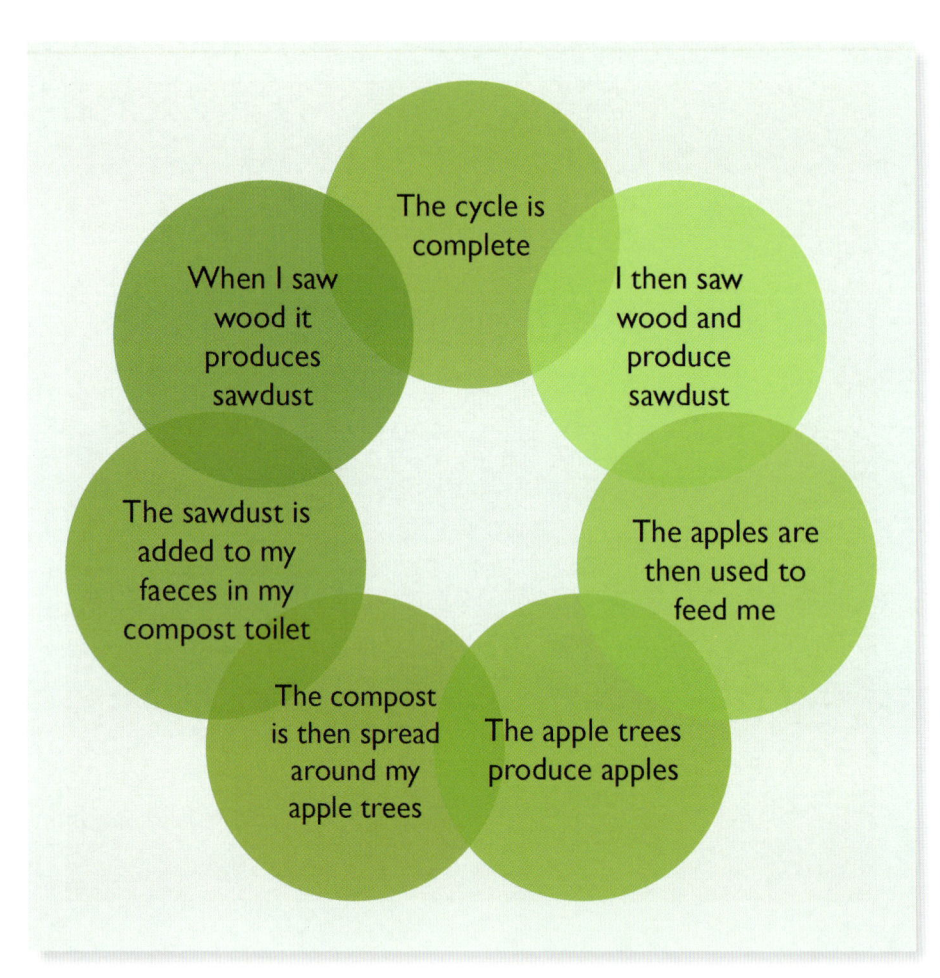

- The cycle is complete
- I then saw wood and produce sawdust
- The apples are then used to feed me
- The apple trees produce apples
- The compost is then spread around my apple trees
- The sawdust is added to my faeces in my compost toilet
- When I saw wood it produces sawdust

Edge Effects, or Using Edges & Valuing The Marginal

Edges are the interface between two media: the shore where the sea meets the land, where the woodland meets the field, where water meets the air. Wherever species, soils, elements, or natural or artificial boundaries meet, edges appear. Edges are highly productive environments. If we take the urban context of the edge between the kerb stone and the road, it is here that leaves will congregate, that paper bags will collect and where children will prefer to walk. If we look at the edge between woodland and grassland, we find that the edge contains both species from the woodland and the grassland, as well as species unique to the edge itself!

The majority of sustainable cultures have understood the benefits of edges and are often sited where they can draw from at least two environments, between land and water or forest and field. Hedgerows are a beautiful example of edge effects, diverse and abundant, trapping wild energies like wind and catching bird manure and spreading seed to surrounding fields.

Edge patterns are nature's way of helping us use space by growing crops or trees in patterns that nature has already identified for us. One example is the spiral pattern, so clearly shown on the shell of a snail. If we create a landscape following the upward spiral, we gain more vertical space, increase surface area and can stack more plants into a small space.

A lobular pattern can be particularly useful when planting into a naturally regenerating woodland. For example, in the pioneer phase when a large area is covered with perhaps broom and gorse, a lobular pattern can be cut to create planting space for trees which will be sheltered by the existing gorse and broom and will create successful clump plantings and utilise space more efficiently than a straight line planting. Sun traps can then be created along the woodland edge by pattern designed planting to utilise this productive space for the growing of crops.

A woodsman like myself lives on the marginal yet fertile edge of society. It is a life rich in its diversity of activities and yields, yet challenges conventional thinking, planning and social convention by representing a sustainable way of living with values based around future generations rather than just our current one.

Design Through Patterns Into Detail

Patterns can be seen throughout my work at Prickly Nut Wood. The seasonal patterns of cutting coppice during the winter, allows the light in and thereby enables spring flowers to grow. The coppice provides craftwork, producing products for gardens in spring. Timber framing is carried out in the better weather of summer, and autumn is for harvesting produce, wild food and fungi to store for the winter.

Species patterns occur in the woodland in the form of crops like the appearance of the fungi Horn of Plenty towards the end of October. The chestnut and apple harvests form their seasonal patterns as does the birch sap wine, tapped in the spring ready to drink in the autumn; and the reverse with cider, pressed in the autumn ready to drink in the spring.

There are also migratory patterns. The nightjar returns every year from Africa to nest in the freshly cut coppice. The male arrives two weeks before the female and marks out his territory by calling with his dusk 'churring' sound. There is an evening every year when I sit out and I sense it is time for the nightjar to arrive. Usually within a day or two, I hear the familiar 'churr'.

There are landscape patterns, formed by generations of coppice workers who have cut particular cants, which future generations follow, realising the wisdom in the size and choice of cant.

Woodlands have formed their own unique set of patterns and as we interact with the woodland so we develop our own patterns in conjunction with those we learn through observation.

Produce No Waste

A natural broadleaf woodland is a closed system. The trees 'waste' is their limbs and leaves that fall to the ground. These are then utilised to feed further trees and are broken down by invertebrates and fungi, all part of the diverse woodland ecosystem. When we interact with the woodland, we must follow this pattern and ensure our wastes can be utilised back into the cycle of woodland management.

Above:
A planted wildlife corridor, Lodsworth, West Sussex.

I try and apply this to the produce from the woods and to the roundwood buildings I construct. I am appalled by the huge amount of waste that is either buried or skipped on a building site. If you are building with roundwood and natural materials, there is no need for a skip. The 'waste' you are left with is a large pile of offcuts of timber. These are not waste but fuel to heat the building you have just constructed. With many modern buildings, we are leaving waste and pollution for the next generation to clear up. By using non-recyclable or non-biodegradeable materials, we are closing our eyes to the waste we leave for the future. A roundwood timber frame building from the foundations to the ridge leaves no pollution for future generations.

Attitudinal Principles
The previous principles are all interrelated principles of designing. They all need to be used in conjunction with each other to design a permaculture woodland. Attitudinal principles are principles that are people principles and involve you, the designer or designers, and are essential if you wish to free up your creativity and design a permacultural woodland.

Everything Works Both Ways – 'The Problem Is The Solution'
It is very easy to become attached to the idea that a particular circumstance is a problem.

We've all done it, but few of us take time to step back and turn the 'problem' around, to see it the other way – as a solution. A few years ago I visited a farm to give some advice as a permaculture consultant. One of the main concerns of the farmer was a boggy wet field. It had lodged in his mind as a problem and he used large amounts of his energy, feeling frustrated with the piece of land and how much it would cost to drain it. I suggested he no longer saw it as a problem but turned it the other way and saw the boggy ground as a benefit, and that it was the ideal place to plant willow. There was a grant available for planting short rotation coppice willow (see Grants in Chapter Eight) and a ready market as there was a short supply of willow in the area. Local basketmakers and sculptors were buying in willow from Somerset. He could also consider using the willow as a fuel crop if he installed a suitable boiler to heat the farm. His face lit up. It was as if a veil of worry had been lifted.

This principle of attitude is essential in the design process and is also useful in all other aspects of our lives.

Yield Is Theoretically Unlimited
Permaculture design is information intensive as opposed to energy or capital intensive. Information represents ideas, knowledge and experience of generations of people before us. The yield of any woodland is not limited by its size, its only limitation is in us as the designer. I can design a woodland using all the knowledge I have and all the information I have collected, but the yield of that design is limited by my mind, another person could look at the design and add new ideas to increase the yield. So having designed your woodland, discuss it with others, share ideas and increase yield.

Position In The Landscape
When planting a new woodland think about the relationship of the new woodland to existing woodlands, hedges and shelter belts and consider designed wildlife corridors to link the new woodland to existing trees. A wildlife corridor can be a hedge, a thin strip of woodland or an area of pasture which is left undisturbed to regenerate.

Woodlands should be planted on steep slopes and inclines to reduce risks of soil erosion. The digging of swales should be considered on all slopes. Design carefully all the access rides, remembering it is less work to bring timber downhill than to take it up! Be content to have areas that are inaccessible, these can be left to nature and will be of great benefit to wildlife. North facing slopes are a good choice for woodland as trees grow successfully without full sunlight and south facing slopes are ideal to terrace for vegetable growing.

Creatively Respond to Change
With climate change, every woodland will undergo significant changes as species reach the edge of their climatic comfort zone. We cannot tell how fast or how effected certain species will be, but we can design and gauge what species may do well in the future in our particular location. We should not just be considering planting a tree species because it may grow well. We should be looking at future uses and needs and how our chosen species will adapt and fill those niches. We will need to experiment, try different species and observe and learn from how well they establish. We will need species for fuel, building materials, and food and species that will be beneficial to wildlife. We will need to phase in new species as others perhaps migrate north or are lost to the woodland through plant pathogens.

Creatively responding to change will affect our management strategies for woods as well. At Prickly Nut Wood, for generations, the sweet chestnut has been grown as a single species coppice with the diversity being the bryophytes below the coppice and flowers along the woodland rides. I am now encouraging birch and alder regeneration by following a coppice cut with another a few years later. This process allows the naturally regenerating birch and alder to not get shaded out by the regrowing chestnut but rather to be drawn up and become part of the coppice woodland. This process of increasing tree diversity within the established coppice is responding to the increased risk of tree diseases and the realisation that a more diverse woodland has a better chance of survival.

Fossil fuel use is a large part of the management of the majority of our woodlands. As change occurs, we need to find alternatives. Horses for extraction and traditional crosscut saws are options from the past that have an important place in the future, as do technological improvements like the battery powered chainsaw. I have just cut three acres of coppice with battery chainsaws, recharging the batteries by stored solar power each evening.

TWO EXAMPLES OF DESIGNED WOODLANDS

GOUDHURST
In 1993, I designed a woodland in Goudhurst, Kent. The site was a 10 acre (4 hectare) arable field that had been a hop yard previously. The wood was planted using permaculture design principles. The first part of the design involved increasing the woodland edge. The edge was greatly extended with the use of 20-23ft (6-7m) wide rides which are cut twice a year to encourage wild flowers. Along these rides fruiting edge species were planted comprising crab apple, wild service, guelder rose, rowan and cherry. Different apple cultivars could then be grafted onto the crab apple once it was established and likewise with varieties of cherry onto the wild cherry (cherries are likely to need bird protection). Harvesting produce is simplified as the fruit is on the edge of the ride where it

Above:
David Scott planting trees at Goudhurst, Kent.

Over page:
Woodland design at Goudhurst.

56 The Woodland Way

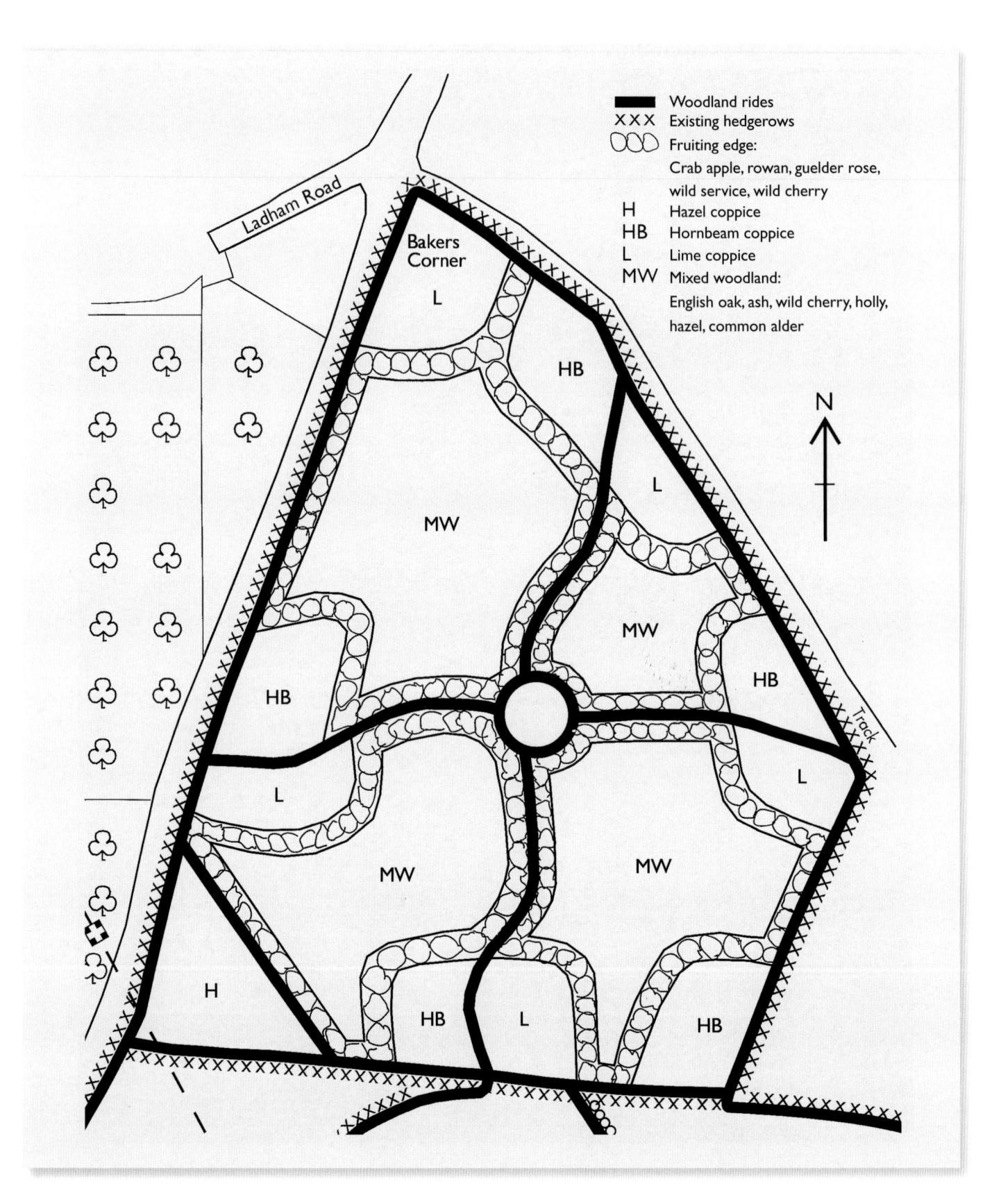

is easy to pick and transport.

Ten coppice blocks, five of hornbeam, five of small leaf lime, were planted. These are all on the corners of the rides, in small blocks of around 80 trees for ease of management. A wide ride will ensure sufficient light for good regrowth after coppicing. Once a block is coppiced, it allows extra light into the fruiting species behind and encourages vigorous fruiting until the coppice regrows. In this way, coppice rotation also extends the fruiting edge. The hornbeam coppice is for use as a firewood and the lime for greenwood turning. The lime also reaches flowering before it is coppiced, ensuring a tasty honey crop should bees be introduced. A hazel coppice was planted under the power lines (in the south western corner) after negotiation with the local electricity supply board. The hazel is harvested on a seven year cycle to ensure that it does not interfere with the power lines and there is presently a high demand for in-cycle hazel. (Since I made this arrangement I have come across coppice workers not being allowed to cut under large national power lines for safety reasons, whereas the lines at Goudhurst were small domestic lines.)

Further away from the edge are high value timber species of which oak (*Quercus robur*) is the dominant species. The rest is made up of ash and wild cherry. Holly has been planted randomly to allow some evergreen cover with hazel producing a shrub layer. Alder has been planted throughout to fix nitrogen and is coppiced where it grows adjacent to the lime and hornbeam blocks. The woodland has been planted with 12 species, and blackthorn and hawthorn will spread from the hedges to increase the shrub layer. A central glade is designed to allow a workshop area to add value to the harvested products.

Top:
Meera's wood staked up, ready for planting. Note swale holding water. Lodsworth, West Sussex.

Below:
Meera's Wood 18 years after planting, now blending into the landscape.

MEERA'S WOOD

In 1992, I designed and planted a new woodland in Lodsworth, West Sussex. This site was 3 acres (1.2 hectares) of sloping pasture with a clay/loam soil. The first stage was ground preparation. I dug swales on the contour to allow extra water infiltration to the young trees. Swales are ditches which rather than channel water, collect it and then allow it to be absorbed into the ground helping the young trees through summer. The slope is about 35 degrees and when it rains on a summer's day, about 80 percent of the rainwater runs down the slope and is collected in the streams at the bottom of the valley. Swales ensure that the majority of that water is collected within the new woodland and fed to the trees.

The summer of 1993 was dry and I lost two trees out of 1,400, an excellent survival rate by any planting standard. I put it down to the swales. The woodland is planted with 14 native broadleafs of which oak (*Quercus robur*) and wild service (*Sorbus torminalis*) are dominant. The long-term management has two possibilities:

1. A mixed coppice with oak standards, coppice materials to be used for craftwork and charcoal burning, oak for high value timber.

2. A mixed coppice of wild service standards, coppice material to be used for craftwork and charcoal burning, wild service to be harvested for berries, sold to make a traditional alcoholic drink and later the timber to be sold to a specialist market.

The second possibility seems the most likely direction as a local buyer is interested in producing the beverage from the berries, but the first option is easy to switch to. Here I have looked ahead to find a new market, but also cover myself with the backup of the traditional woodland system. The wood from wild service is hard to find and fetches the highest price of native woods in France. It is a specialised market often favoured by musical instrument makers and so to sell some standard service trees is also likely to be a useful economic addition to the design.

This current winter, I have been through the now well established 19 year-old woodland and marked up some of the poorer quality trees ready to thin which are of a size to produce a

Above:
The author cleaving chestnut pales. These make excellent tree stakes.

58   The Woodland Way

firewood crop. The wild service trees have done well in some areas but not so well in others, so I anticipate developing a mixed coppice with standards with wild service and oak standards.

Hence the management strategy has amalgamated the two possibilities I foresaw at the time of planting and has highlighted the 'apply self regulation and accept feedback' principle. The woodland has developed and it has chosen where certain trees have done better than others. I am accepting that feedback and in doing so the woodland is becoming more diverse than I had originally designed. The mixed coppice will have a standards layer producing the wild service berry and some specialist timber as well as oak standards benefiting over 300 species of invertebrates and over time producing a high value timber crop.

Planting Trees & Aftercare

Trees are wonderful survivors and despite the poor quality of many soils, they will do their utmost to grow. Provided basic care is taken at planting time, results should be good. Sourcing good seed stock from a local nursery which is concerned with local provenance of seed will ensure the trees you have planted are adapted to the area and that there will be less risk of hybridisation due to crossbreeding from foreign seed stock. Local seed stock produces trees which are adapted to local soil conditions and climate and therefore is the first preference for selection. Local stock also keeps the ecology in balance and holds on to the character of local distinctiveness. If you are ordering from a local nursery, try to give them at least six months notice. This is also your commitment that you will plant, and gives you plenty of time to prepare. Even better than a local nursery is to propagate your own seed, then you will have the full history and origin of the trees that you are planting. If sourcing your own seed, look at the seed parent tree for vigour, health and characteristics you would like to see in the trees you will be growing. For this you will either need to create a small nursery, or with certain varieties that germinate from a nut, you might consider direct seeding. Whatever you decide, you should have the ground prepared well in advance. On pasture, a grazing of sheep prior to planting will ensure the vegetation is cut low and has been recently manured. If the ground has been heavily compacted by machinery, it might be worth considering preparing the ground with a subsoiler (see Reclaiming Degraded Land later in this Chapter), always cultivating on the contour.

If tree shelters are needed they should be ordered well in advance and stakes cut and prepared. Sweet chestnut is a recommended species for shelter stakes as they are durable, obtained from coppice wood, and up to eight stakes can be cleft from a 5ft 6in (1.68m) length of chestnut cut on a 12 year rotation. You will often get at least three 5ft 6in (1.68m) lengths producing 24 stakes from one coppiced pole! Any fencing should have been finished in advance of planting and a heeling in bed prepared for when the trees arrive. Many people advocate drainage of boggy grounds prior to tree planting. To me this is a waste of resources and energy (see Attitudinal Principles earlier in this chapter). If the ground is boggy, plant trees that want to grow there, like willows and alders.

The main cause of loss of young trees is the drying of the roots caused by exposure to the wind and sun. When collecting bare root trees be sure to have the roots enclosed in a windproof bag and heel them into the ground as soon as possible, unless you are planning to start planting that day. When taking trees from a heeling in bed, take no more than a hundred at a time, keep them in the bag until the hole is dug and move them from the bag to the ground as quickly as possible. Never lay a tree on the ground while you dig a hole. Do not plant on windy days or when the ground is frozen. Broadleaf trees are best planted during November and December. This gives them the rest of winter to settle in and their chances of surviving the first summer are greatly increased. Evergreens are best planted in early spring to avoid the late frosts. Notch planting is the quickest and simplest planting method and the most suitable for planting small trees. A Mansfield planting spade is a worthwhile investment and is designed for this type of planting. To notch plant, scrape away any vegetation and dig the spade into the ground to open up a deep slit into which the bare rooted tree is placed. Be sure that the slit is deep enough for the roots and once placed that the notch is firmly closed up with your foot – especially on clay soil as they can reopen in dry weather. Planting trees of about 12in (30cm) in height is ideal with this method. This method is for young trees. Larger trees such as fruit trees will need a pit dug and compost added. Remember the best way to plant a tree is from seed! The larger the tree, the longer it takes to stabilise in the ground and the more labour intensive it is to plant.

Once planted, the control of grass and competing vegetation is vital in order that the tree gets off to a good start. Commercial herbicides are the most commonly used method but if, like me, you prefer to find other methods, then mulching is the preferred choice. There are commercially available mulch

Species	Tree age when good seed mast produced	Collection time	S: Sowing ST: Stratify	Comments
Ash	25-30	Aug-Nov	S: Mar-Apr ST: 17 weeks	Seed collected in August can be sown directly
Beech	50-60	Sept-Nov	S: Early Mar ST: 3-4 weeks	Can be sown immediately
Birch	12-18	Aug-Sept	S: Mar-Apr	Collect catkins before fully ripe
Common alder	15-20	Sept-Spring	S: Mar-Apr	Pick when cones start opening
Corsican pine	25-30	Dec-Feb	S: Mar-Apr	
European larch	15-20	Dec-Feb	S: Mar-Apr	
Douglas fir	30-35	Oct-Nov	S: late Feb-Mar ST: 4 weeks	
Field maple	25-30	Sept-Nov	S: Mar-Apr ST: 18 months	
Hazel	12	Sept-Oct	S: early Apr ST: 3 months	
Hawthorn	8-12	Sept-Nov	S: Mar-Apr ST: 16 months	
Holly	18-25	Nov-Jan	S: Mar-Apr ST: 16 months	
Hornbeam	25-30	Aug-Dec	S: Mar-Apr ST: 12-15 months	
Oak (Q. robur & petraea)	40-50	Sept-Nov	S: Sept-Nov	Sow immediately after collection
Scots pine	15-20	Nov-Feb	S: Mar-Apr	Collect cones before fully ripe
Small leaved lime	20-30	Sept-Oct	S: Mar-Apr ST: 18 months	Propagation by root cuttings easier
Spindle	15	Sept-Nov	S: Mar-Apr ST: 4-6 months	
Sweet chestnut	30-40	Sept-Oct	S: Sept-Oct	Sow immediately
Sycamore	20-30	Sept-Oct	S: April ST: 1 month	Can be sown immediately
Walnut	30-40	Aug-Sept	S: Mar-April ST: 6 months	
Whitebeam	8-12		S: Mar-Apr ST: 6 months	
Wild cherry	15-25	July-Sept	S: Mar-Apr ST: 13 weeks	Can be sown immediately
Yew		Sept-Nov	S: Mar-Apr ST: 16 weeks	

mats, which suppress the growth of competing vegetation and help to warm up the soil in spring, but they are not cheap. I would only recommend them when you are planting hundreds of trees rather than thousands. Mulching on a smaller scale can be done with waste materials; carpets and cardboard are very effective, but must be well secured. It is essential to ensure that the ground is moist prior to mulching (which it usually is in winter) as mulching will hold that moisture, but can also keep a dryer patch of ground dry.

Another useful suggestion is to bring some leaf mulch from an adjacent wood and spread it amongst the new trees to start off the necessary fungal relationships on which the woodlands are so dependent.

Creating A Tree Nursery

If you are involved with working in woodlands and notice the hundreds of young trees that can germinate after a heavy mast year under one mature oak or beech, it will no doubt occur to you at some point to start a tree nursery. A tree nursery will provide local provenance stock as the seed can be collected from local woodlands. It will give you all the replanting stock you need and surplus to sell or plant new woodlands with.

A nursery consists of a seed bed, where germination takes place, and other beds for transplanting the seedlings for growing on. These beds are best slightly raised and a loamy soil is preferable as it is easier to work than a heavy soil and doesn't dry out as fast as lighter sandy soils. These beds must be enclosed in a rabbit and deer proof enclosure and have easy access to water and some form of shading. For seedlings in high summer, bracken can be used successfully as shading. The germination of tree seed depends upon the variety.

Right:
One year old hazel establishing in a woodland nursery bed.

Seeds are sown in the seed bed as densely as possible.

Small seeds should be pressed into the soil with a wooden board and covered with a fine soil. Water regularly during spring and summer and don't allow the soil to dry out at any time. Depending on the variety, seedlings will be moved from a seed bed after one or two years and transplanted into a transplant bed. The process of transplanting encourages vigour and some nurseries 'undercut' the roots to create a similar effect. If you buy young trees from a nursery (or if you sell from your own) they will be described by time spent in the seed bed and time spent in the transplant bed. So a young tree described as 1+2 means a three year old tree of which one year is spent in the seed bed and two years in the transplant bed. You may also come across 1 U 2. The U stands for undercut and this seedling will be three years old, one season before undercutting and two since.

Some species are slow to germinate, so be patient if you have not stratified the seed.

The fertility of the seed and transplant beds will need to be improved as each batch of trees will take out many nutrients. Organic liquid fertilisers made from nettles and comfrey as well as intercropping the beds with green manures like lupins and field beans will ensure that the beds remain fertile and topping up with compost will allow for soil loss when the trees are moved.

Green Manures In Nursery Beds

Broadcast green manure seed over the seed bed and allow to grow until just before the green manure is maturing (i.e. before flowering) and then dig the

Plant	Annual Perennial Biennial	Bee plant	Soil	Nursery bed, Woodland establishment	Sowing time	Digging in time	Fixes nitrogen
Field beans	A		heavy, not dry	NB	Sept-Nov	up to flowering	✓
Buckwheat	A	✓	tolerates poor soil	NB	Aug-Apr	up to flowering	✗
Clover, Alsike	P	✓✓	tolerates damp,	WE	Apr-Aug	any time	✓
Lupin, bitter	A	✓	light, slightly	NB / WE	Apr-Jun	up to flowering	✓
Phacelia	A	✓✓	most	NB	Mar-Sept	before flowering	✗
Rye, grazing	A		most	WE	Aug-Nov	before flowering	✗
Tares, winter	A		heavy soils	WE	Mar-Sept	up to flowering	✓
Trefoil	A/B	✓	light soil, not acid	WE	Mar-Aug	Any time	✓

Establishing New Woodlands

whole plant into the soil. The bed should be ready for planting three to four weeks after digging in.

All are hardy except buckwheat, (very tender), and phacelia which will only survive mild winters. Trefoil is the most recommended for woodland establishment as it will survive in partial shade.

Stratification

In woodland conditions, seeds will remain in the ground for varying periods of time in a state of dormancy before germination. In a nursery, this can be imitated by constructing a pit, and placing the seeds within it. This process is called stratification. The pit should be well drained and filled with 16in (40cm) of sand with the seed planted at a depth of 8in (20cm). The pit should be exposed to seasonal rainfall, but in a shady place to avoid the sun heating up the top part of the pit and causing early germination.

Potential damaging species	Height of protection needed
horses	2.50m
cattle	2.25m
red deer	2.0m
goats	1.85m
fallow deer	1.8m
roe deer	1.2m
sheep	1.2m
rabbits	0.60m
voles	0.20m

Bare Root or Container Grown Trees

I prefer bare root to container grown trees. Container grown trees give a certain 'laziness' to the planter whereas the naked sight of a bare root tree encourages the planter not to hang around!

Container grown trees in theory can be planted at any time of year, but I do not recommend planting except during the dormant season. Root trainers are the best containers for seedling trees encouraging the tree to develop a straight root system. This should start them off well when transplanted into open ground or into a woodland situation. Evergreens such as holly, yew or Scots pines are best obtained as root trained where possible, although all of them germinate well from seed.

Tree Shelters

Tree shelters are used with increasing regularity and here I have outlined some positive and negative features of them from my experience.

Positive

Protection. Depending on the height of the shelter they give excellent protection from rabbits and deer.

Growth. They encourage fast growth of trees: wild cherry, alder and rowan get away particularly well.

Warmth. The shelters act as mini-greenhouses, getting the trees off to an early start in spring.

Visibility. The trees are easy to find, even in winter.

Straightness. The tree grows up straighter towards the light, hopefully producing a higher value timber.

Fencing. You don't have the expense and time of erecting rabbit and deer fences.

Agroforestry. It is possible to graze geese and small stock on the land while the woodland is establishing.

Negative

Cost. They are expensive, costing often three times as much as the tree, plus you will need a stake with most types.

Plastic. They are made from plastic! It is supposed to be solar biodegradable but I find that the canopy closes, blocking out sun before they biodegrade. (However this is useful as you can cut them off and reuse them!)

Visual. Just because we can see a field with tree shelters, doesn't mean the trees are doing well inside and quite often they tend to be neglected.

Wildlife. Occasionally small birds slip down the shelter and get trapped.

Ventilation. Because the shelters act like a mini-greenhouse, trees can get overheated and some species like beech, in particular, suffer. I always pre-drill ventilation holes near the base of the shelters – but this should come as a standard.

Wind. If they are not well staked and get blown, the tree will grow at the angle of the shelter.

New. Tree shelters were first tried in 1979, and so we do not know the long-term effect of timber quality with the fast growth rate in the early years. In 50 years time we will have a better understanding.

My opinion at this time is that shelters are very useful in certain situations. In particular, for restocking areas of existing woodland where fencing would be impracticable and not economically viable, and also for use in agroforestry systems where livestock will be grazed beneath the trees, but I would not be inclined to use them for a broadscale new planting. Fencing will usually be a cheaper alternative on a large-scale planting unless the planting is of a very irregular shape with lots of edge, like a shelterbelt where tree shelters may prove a cheaper option.

If you are planting small numbers of trees, consider making your own tree shelters or using second-hand tree guards as the biodegrading process of the guards allows them to be used more than once. Cut them off the existing tree and then tie with string when reusing.

Plant Spacing

The Forestry Commission recommend the planting of 910 trees per acre/6ft10in x 6ft10in spacings (2,250 trees per hectare/2.1 x 2.1m spacings), for both broadleaf and coniferous plantings. You need to keep to these numbers to obtain planting grants but you are allowed up to 20 percent open ground. When creating new native woodlands, (as in the aforementioned Forestry Commission Bulletin No. 112, 'Creating New Native Woodlands'), amenity use woodlands, and small broadleaf woodlands of less than 7.4 acres (3 hectares), a planting distance of 9ft10in x 9ft10in (3m x 3m) is allowed by the Forestry Commission. This works out at 445 trees per acre (1,100 per hectare). If you are not planting under the grant scheme then the choice of planting distance is up to you. The key point to remember is that closely planted trees will draw each other up towards the light and will tend to produce straight timber of higher value, if timber is your reason for planting. The further apart the trees are planted, the more room they have to stretch out and find their individual shape. They will be of low value for timber but of higher value for recreation, physical beauty and will have many wildlife benefits like more branches for nesting sites.

Planting Patterns

Planting patterns will be dependent on your long-term aim for the woodland. A newly planted woodland which will become coppice, will be densely planted with 5ft9in-6ft10in (1.75-2.1m) spacings for hazel/sweet chestnut on fertile soil, whereas a recreational woodland may consist of clump planting in open spaces. Clumps may be of one species or designed as species groups where species have symbiotic relationships. Numbers of trees in clumps will be dependent on the whole design of the woodland, those trees in the centre of the clump will be drawn up and those to the outside will branch outwards. Internal spacing within the clump will depend on the management planned for the woodland. It is also possible to plant purely at random, but clump planting is generally more successful as slower growing species are not shaded out. Close planting within clumps of 6ft6in (2m) apart or less should be used if timber quality is the main importance, since wider spaces will allow more early branching of trees.

Aftercare

If the trees have been well mulched, competing vegetation will be kept at a distance and the trees should get away well. Grasses are best left wild and cut once a year as a hay crop until the canopy closes and they start to die back. If cut regularly or mown, they are stimulated and grow more, taking more nutrients and water from the soil.

Beating Up

'Beating up' is a slightly bizarre forestry term used for inspecting trees and replacing any that have failed to survive. Beating up takes place the winter after planting. Dead trees should be marked in the summer if you are not used to identifying them in winter. While beating up check all the other trees and firm the ground around any trees where the ground has opened following planting. If tree shelters have been used, check the ties that secure the shelter to the stake for tautness, stakes for stability and the straightness of the tree shelter. Deer and rabbit fences should be checked for damage. (If the woodland has been well planned, the ground prepared, good tree stock sourced and providing care was taken with planting, beating up should not take up much time.)

Pruning

'To prune or not to prune'. I try to avoid the need for pruning by utilizing good stocking rates to draw up the trees and naturally allowing the side branches to die off as they are shaded out by the other trees. When deciding whether to prune, you must weigh up the time and labour involved in pruning against the benefits to the tree and, in the long-term, timber. If the goal of the woodland is to produce high value timber, then the removal of side branches up the stem will help produce a notch free timber. In conifer species which are used for joinery, this will help with stress grading

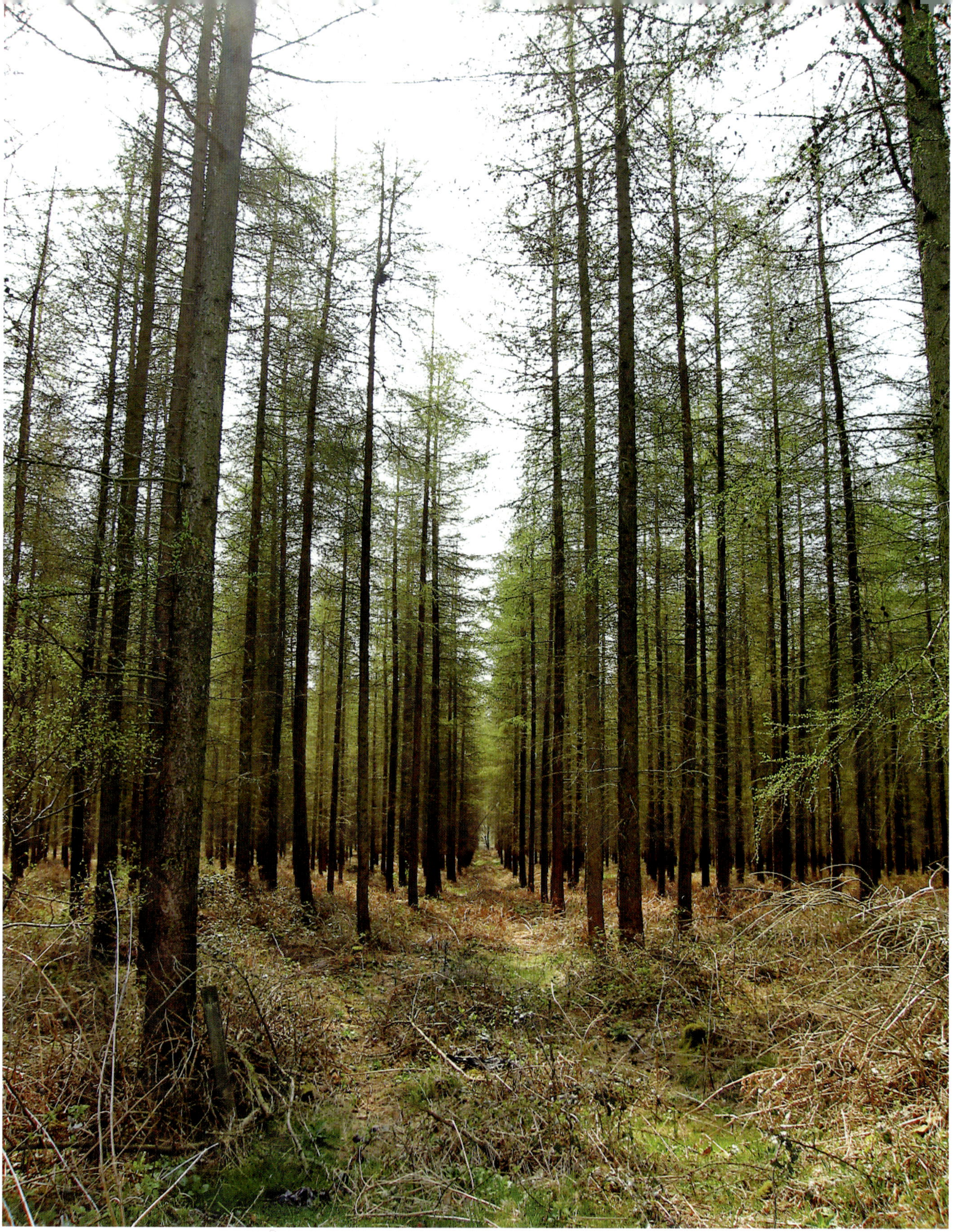

and, in the longer term, building regulations. Other times when pruning may be necessary is to remove a fork in the leader stem so it grows on straight as a single stem tree and also for the removal of any diseased branches. Timing of pruning is important to ensure disease does not occur. For example, wild cherry is best pruned in July to avoid risk of silverleaf infection. Most pruning is carried out over the first twenty years of a tree's life and is often done in stages, removing more branches as the tree grows upwards. At least 30 percent of the branches and top should be left on the tree so as not to reduce vigour. Closely planted woodlands or a dense patch of naturally regenerating woodland will, to a degree, self-prune as many side branches will get shaded out at a young age and then drop off. If we let nature do the pruning of our woodlands for us we can direct our pruning energies to orchards and encourage the increased yields of our fruit trees.

If you do wish to prune for high value timber, Cyril Hart's *Practical Forestry for the Agent and Surveyor* has some detail on different species and times to prune (see Bibliography in Appendix Seven).

Invasive Species

Invasive species will need to be controlled when establishing new woodlands as well as when managing existing ones. Pioneer species will often naturally regenerate in amongst a planted woodland. For example, species like birch, which will help draw up the planted trees, are a benefit and can be thinned out at a later stage. Wooded shrub species like rhododendron and laurel will become a challenge in any woodland if left to grow unchecked. Rhododendron has in the past been planted in woodlands to allow cover for pheasants. These are often old coppice woodlands where shooting was the main use of the woodland. The rhododendron eventually spreads, covering the ground with a dark shade, stopping any natural regeneration and acidifying the soil as its leaves break down. As the old coppice stools start to collapse, the woodland eventually becomes pure rhododendron. Control of rhododendron is a laborious task.

Left:
Recently pruned larch plantation, Fernhurst, West Sussex.

Above:
Rhododendron ponticum
well established in woodland, West Sussex.

It both freely self seeds and suckers and cut branches laying amongst damp leaves on the woodland floor will easily root! The commonest recommendation is to cut the rhododendron to ground level and then spray the new regrowth with a herbicide. (This is possibly the only method that really works for total eradication and will take more than one application.) I have had some success using a thick lime solution which is painted onto the leaves. Rhododendron is an acid loving plant so feeding it lime can eventually kill the plant, but it has taken me many applications over about three years to succeed! The other option is to winch the roots out, but this is a time consuming activity if the rhododendron is well established and can also disturb a lot of the ground flora in the process. Woodland Improvement Grants are available from the Forestry Commission to remove rhododendron from woodlands. I have sold some of the small rhododendron for kindling, the larger wood makes interesting rustic furniture and excellent firewood. I used to think we could manage the rhododendron by continual harvesting, but now it is a known host plant for the pathogen Phytophora and we have no option but to put in many hours of hard work to eradicate it. Other potential invasive plants in woodlands which need controlling are Japanese knotweed and Himalayan balsalm.

Reclaiming Degraded Land

By degraded land I am referring to land which has suffered from soil erosion: old quarries and spoil heaps, demolition sites in urban areas, and land which has been degraded by the use of agrochemicals.

When examining degraded land, the source of the degradation must be established and stopped. In the case of soil erosion, retaining walls and the piping away of rainwater may be necessary to control further soil loss in the early stages of reforestation. A ground cover layer may be needed to stabilise the soil prior to tree establishment. A grass sward with a high proportion of legumes (clover or trefoil) is recommended.

The soil structure should be examined as compacted soil is a common problem on degraded land and is generally caused by the continual use of heavy machinery. Compaction will affect root penetration of young trees, can reduce the water holding capacity and reduce the aeration of the soil. The ground may need to be prepared by using a soil reconditioning unit or by cultivating with a subsoiler with large-winged tines to a depth of 18in-2ft (46-61cm). Cultivation should take place on contour (to encourage water infiltration and avoid soil erosion) when the soil is dry enough to fracture but not so dry as to create large clods.

Some degraded lands will have little or no top soil available. Although many trees will establish on subsoil, provided another source of nutrient is supplied, the lack of top soil often leaves any newly planted trees deficient in nitrogen. Species which fix nitrogen like alder and acacia should therefore form a large part of any new planting.

Many degraded soils are contaminated. Some contaminants have little effect on tree growth but others, heavy metals in particular, have adverse effects on most vegetation. An estimate for 'potentially contaminated land' in the UK, was thought to be about 67,00 acres (27,000 hectares) in 1989 (Forestry Commission Bulletin No. 110). If contaminated land is suspected, laboratory analysis should be undertaken to discover the nature of the contamination and whether the land is hazardous to the health of trees or to that of people and wildlife who might venture onto it. For an analysis, contact the Institute of Professional Soil Scientists (see Useful Organisations in Appendix Eight).

Having established the nature of contamination, guidelines laid down by the Interdepartmental Committee on the New Development of Contaminated Land should clarify whether prior action to tree planting is needed or not. These guidelines are available from the Department for Environment, Food and Rural Affairs, reference guidance note 59/83, second edition.

Tree planting on degraded land should be carried out as for a new woodland planting on arable land, using notch planting for broadleafs of 1ft4in-2ft (40-60cm) height and conifers of about 8in (20cm). Pit planting should be used for heavy clay soils and container grown plants.

The most recommended broadleafs for degraded land are primarily nitrogen fixers, so all varieties of alder, acacia and false acacia are recommended. Willow species are useful for ease of establishment and for their good root systems for soil stabilisation. Other broadleafs that have been successfully used are grey poplar, birch, field maple, white beam, sycamore, rowan, hawthorn and, to a lesser extent, oak. The most successful coniferous species is Corsican pine.

A grant is available for the reclamation of derelict land. The Derelict Land Grant (DLG) is given to land which in its present condition reduces the attractiveness of an area to live in or work in or, because of contamination, the land is a threat to public health and safety. The grant is available from the Department for Environment, Food and Rural Affairs. The Derelict Land Grant may pay for prior work such as soil analysis for contaminated land and studies and drawing up plans before any actual planting is undertaken. It is recommended to investigate the possibilities of grant aid prior to carrying out any work on a degraded land site. (Further reading: Forestry Commission Bulletin No. 110, 'Reclaiming Degraded Land For Forestry'.)

New Planting Initiatives

The National Forest

The National Forest was launched in 1990. Its objective is to achieve a multi-purpose forest of over 200 square miles (518 square kilometres) in the Midlands of England. The forest sits in the middle of an approximate rectangle formed by Birmingham, Stafford, Nottingham and Leicester. About 30 percent of this forest area will actually become working woodland with the planting of 30 million new trees, over 33,400 acres (13,500 hectares). The programme is run by the National Forest Company (see Useful Organisations in Appendix Eight).

Community Forests

The 12 planned community forests are situated close to towns and cities. Presently existing tree cover in community forest areas is about 6.9 percent, below the national average; the aim is to increase tree cover to about 30 percent. Community forests are planned to be diverse forests catering for the needs of those within the areas of the community forests. Local people are encouraged to take an active role in planning and using the benefits of the community forests. (See Community Forests in Appendix Eight for contact details for the 12 community forests.)

Heartwood Forest

Near St. Albans, the Woodland Trust are currently establishing a new community woodland, Heartwood Forest. The trust aims to plant 600,000 trees over 858 acres of land utilising volunteers, including open spaces and a community orchard. (see Woodland Trust in Appendix Eight.)

Above:
Community tree planting at Pentiddy Woods, Cornwall.

MANAGEMENT OF WOODLANDS

Chapter Five

Establishing Coppice

Coppicing can begin as soon as trees are comfortably established, usually at about seven years of age, but it is worth considering what you want to do with the wood from the first cut allowing it to grow to a size which is of use to you before cutting it. When coppicing a young tree it is worth ensuring the first cut takes place in late March to early April to ensure that the regrowth establishes well after any frosts.

Buying A Standing Coppice

The majority of coppice workers do not own or lease woodland. They buy the standing timber by the acre. Traditionally, large estates and areas where coppice grows in quantity would have annual auctions. The buyer has the opportunity to view the available coppice and bids at an auction for the area they wish to buy and cut. The prices currently fluctuate between about £50 per acre (£124 per hectare) and £350 per acre (£865 per hectare) depending on quality. However, at the Wessex Coppice Group auction in 1998, grade one hazel coppice fetched nearly £1,000 per acre (£2,470 per hectare) with an expected added value return of at least ten times that amount. Overstood coppice usually only has a firewood, rustic furniture or charcoal purpose but with the benefit of a Woodland Improvement Grant (see Chapter Eight) the viability of cutting is greatly increased. With overstood hazel, hurdle rods, thatching spars, bean poles and pea sticks may also be available in small amounts as well as firewood, walking sticks, rustic furniture and charcoal.

If you are buying a standing coppice, look in detail at the access to the wood and any restriction on access times.

Top:
Recently felled cant of sweet chestnut coppice, Lodsworth, West Sussex.

Below:
Recently felled cant of sweet chestnut coppice, having been converted to craft produce, Prickly Nut Wood. Note that low cutting of stools is usually the best practice to ensure straighter regrowth, but the stools here are cut high to allow more surface area for mosses, as appropriate for this SSSI.

Hazel tends now to be auctioned in three grades introduced by the *Wessex Coppice Group*. These are as follows:

Stools/acre	Rods/stool	Grade
400	30	1
400	20	2
400	10	3
550	22	1
550	15	2
550	8	3
750	16	1
750	11	2
750	5	3

Many estates have shoots and don't like coppice workers around at particular times of the year, although coppicing produces good cover for shoots! Make sure you enter into a written contract agreement (see Contracts in Chapter Eight) with the land owner or estate before starting any work. It is now more common to buy standing coppice outside of auctions by approaching estates direct, or through coppice groups, or the very useful wood selling online magazine, *Woodlots* (see Bibliography in Appendix Seven).

COPPICE MANAGEMENT

Coppice should always be cut in cants and not as individual stools. The area should not be less than 0.3 acres (0.12 hectares) and preferably not much more than a 3 acre (1.2 hectare) block. The exact size to cut will be determined by the overall size of the woodland and access rides. When cutting coppice, smaller species and dead wood (some standing deadwood should be left for invertebrate habitat) should be taken out first to create space before felling larger trees.

Make sure the working area is clear and plan the order and direction in which you are going to fell. Don't fell on a windy day and stop at regular intervals to clean up and sort through what you have cut. When cutting, it is good practice to slope the cut away from the centre of the coppice stool, so that rain water runs off rather than settles on the centre of the stool which will encourage the stool to rot.

A tidy working area is a safe working area. Traditionally, the 'lop and top' or brash is burnt in most coppices. Before burning, consider other uses for the brash: deer hedging, faggots, pea sticks, walking sticks and rustic furniture material, let alone kindling, can often be salvaged from the brash. The important considerations are to remove the brash from the ground so it is not shading out potential flora in the spring and that it is not left in a situation where it could be a fire risk when it dries out or create a habitat for rabbits. The cut stems should be snedded (side branches removed) and cut to product length, if you are certain of your finished product. I leave my poles

Below:
Hazel coppice re-establishing having been overstood for a number of years.

72   The Woodland Way

Opposite page
Top left:
Hornbeam coppice, King's Wood, Wye, Kent.

Centre left:
Small stems are thinned out prior to felling larger stems of 18 year-old sweet chestnut coppice, Prickly Nut Wood.

Right:
Felling the last stem of the cant, Prickly Nut Wood.

Bottom left:
The regrowth begins again – new shoots which may one day become buildings.

Bottom right:
The regrowth now forming into stems, Prickly Nut Wood.

This page
Right:
Claire Godden snedding up chestnut, Prickly Nut Wood.

at the maximum length and stack them so that they can be easily carried out to the nearest ride without damaging any regrowth from the stools. With large diameter sweet chestnut I sometimes leave the wood stacked within the copse for a year to season as chestnut has limited ground flora and the poles will be a lot lighter the following year. I also keep them long so I can cut to order as I get orders for poles from 2ft (0.6m) lengths to 20ft (6m) lengths. Stacked wood should be placed on bearers and preferably be staked at each end to avoid any risk of the stack rolling. It is important where possible to leave standards, standing dead wood and some dead wood on the ground for habitat creation.

Deer, Rabbits, Squirrels & Coppice

Browsing primarily by deer, but in some areas rabbits and/or hares, causes more damage to coppiced woodland than anything else. Deer numbers are increasing and unless we decide that no coppice should be cut, without protective measures being taken against deer, we will end up with a lot of woodland full of nibbled, dying stools. The best way to control deer is through culling by a qualified stalker. By this I don't mean killing all the deer, just keeping their numbers at a level which will not destroy the woodland. Fencing is another option and there have been grants (like the Butterfly Challenge) for specific areas which cover the cost of a 7ft (2.1m) deer fence. Personally, the sight of a 7ft (2.1m) metal fence around a coppice is not a sight I enjoy but it is preferable to seeing all the new coppice regrowth eaten to the ground. Deer fencing is grant aided under the Forestry Commission's Woodland Improvement Grant and there are other options besides a permanent metal fence. Tenax is a cheaper alternative, a black plastic mesh fence that can be tied around trees. It is effective but can be eaten through by rabbits and hares. Heras fencing panels are more commonly seen around building sites but although ugly in a wood, they make an effective temporary deer fence and can be moved onto the next cant once the regrowth is above deer grazing height. I have also experimented with electric fencing, using the tape that is sold for horses. This has been successful but needs regular monitoring to ensure vegetation is not touching the fence and shorting it out. The key with all fencing is to check it regularly, remember a deer trapped inside a fenced area for a few days will cause large amounts of damage.

Using brash is a technique that can work. It is often laid over the cut stool. Although this can deter fallow and roe deer, it is unlikely to deter muntjac and often encourages rabbits, as well as sometimes distorting the base of the new stems. I have heard of people laying large areas of brash at the edge of the copse, sometimes 15ft (4.5m) across, this will indeed deter deer but again, it creates a perfect habitat for rabbits. I have experimented with different forms of brash hedge and decided upon a narrow upright hedge, which uses quite a number of stakes but it is a small sacrifice as it saves the future of the coppice. The hedge is no more that 1ft (0.3m) wide and 5-7ft (1.5-2.1m) at the top, it is

Top:
A steel deer fence ensures the hazel regrowth is safe from browsing.

Below:
Brash hedge under construction, Prickly Nut Wood.

packed densely to deter muntjac as well as roe and fallow deer and is not wide enough to create a safe haven for rabbits to start a warren. If making a brash hedge, it is good to cut some coppice in early October when the leaves are still on, as the leaves will act as a visual barrier and help discourage the deer from trying to jump the hedge.

The brash is laid at an angle and compacted down with spiky tops sticking up as an extra deterrent. The brash hedge must totally surround any area of freshly cut coppice having ensured no deer are fenced in. The hedges need only hold out for two to three years, after that the coppice should have got safely away. Hedges often last longer than this and also make good vertical growing spaces for blackberries. Brambles themselves, despised by many foresters, are a natural protection from deer and soon get shaded out when the canopy closes.

Dung from lions and other big cats can be useful in deterring deer but it is not readily available for most of us and washes away in time. There are chemical repellents which some people say are of use but I have never used them. I have even heard mention of a sensor water gun which fires a jet of water at a moving object, but there are many moving objects in a wood, especially on a windy night.

For me the best and most traditional control is ourselves. In the past, coppice woodlands were full of charcoal burners and chair bodgers who lived and worked in the woods. To cut an area of coppice during the winter and then work on adding value to the wood the following summer is the best protection against deer. I have lived in the woods for the last 20 years and I don't have a deer problem. This applies to a lesser degree to rabbits, but I would advise ferreting of any rabbit warrens over winter to get the numbers down before spring unless the coppice worker is a very keen hunter.

The grey squirrel, *Sciurus carolinensis*, was introduced from North America between 1876 and 1930. Since then they have greatly increased in numbers and are now the dominant squirrel species in the UK, having driven the native red squirrel, *Sciurus vulgaris*, to the fringes of our landscape through dominant territorial behaviour and the spread of squirrelpox virus which is lethal only to the red squirrel. The grey squirrel causes considerable damage to trees through ringbarking the stems, causing the tree growth above the ringbarking to die. In the chestnut coppices where I work, the damage is most severe when

Above:
A good day's squirrel control ensures a good meal.

Management Of Woodlands 75

the regrowth on the coppice stool is 3-4 years of age. In mid summer the grey squirrel will strip the bark of the coppice regrowth searching for the sugars in the sap to feed on. I have had in the past 50% of stems damaged in an acre of regrowth and have since controlled the number of grey squirrels. I do this by roding the dreys (nest made of twigs, leaves and moss) using telescopic lofting poles. This removes their nesting sites and discourages breeding. I then shoot regularly over the winter period using an air rifle. An air rifle with a silencer is far more stealthy than a shot gun and leaves the squirrel meat unspoiled, whereas a shotgun, although effective in killing the squirrel, spoils the meat for consumption. The success of my squirrel hunting has been greatly improved by my lurcher, who points and also catches squirrels when they land on the ground. During the summer, I turn to cage traps, as the leaf cover make shooting ineffective. There is a large responsibility that goes with trapping, which involves regular inspection of the traps. Throughout the year, I keep the numbers in balance and enjoy a number of squirrel dinners for my troubles.

Overstood, Stored or Neglected Coppice

Having cut and protected an overstood cant of coppice, the first regrowth is likely to be poor in numbers of shoots, linked to the fact that stool density may already be poor and some stools may die in the recoppicing process. Some restocking will therefore most likely be necessary. It is possible to use either existing sun shoots or existing older

Above:
Neglected hazel coppice, Milland, West Sussex.

Top left:
Cant of overstood mixed coppice with standards.

Top right:
Same cant freshly felled.

Centre left:
Same cant following spring.

Centre right:
Same cant four years later.

Left:
Lodsworth Larder community village shop built from the wood from the overstood cant.

Management Of Woodlands 77

stems from the overstood coppice to layer and increase the stocking density. Alternatively, vigorous new growth after the stools have been recoppiced can be layered the first winter after the initial cut. All the stools should be cut again after three to four years (second cut) to allow the newly layered stools to be drawn up with the rods from the older stools. The second cut ensures a greater density of new rods than after the initial cut. This means that when the coppice comes back into cycle the quality of the area of woodland is greatly improved. With overstood hazel the second cut is best made three years after the first (which can give a crop of pea sticks or herbaceous border supports) – five to seven years later it is on the rotation for most traditional hazel products.

Layering

There are two possibilities for layering:

1. Leave one or two smaller stems on a stool when cutting, then lay each as you would for laying a hedge, but lay the stems horizontal on the ground. Clear vegetation so that the stem lays on bare soil and remove a little bark where the stem meets the soil at the point you would like a new stool to form. Peg the stem to the ground and cover over with soil where you wish the new stool to grow. Make sure the tip of the stem you have laid is uphill from the original coppice stool, as sap flows uphill.

2. Take a vigorous stem from the first year's regrowth and then follow the steps as above. After a year or two they should have rooted. Cut all the coppice again after three years and the layered stems will be drawn up with the old stools and you will have a good stand of coppice forming.

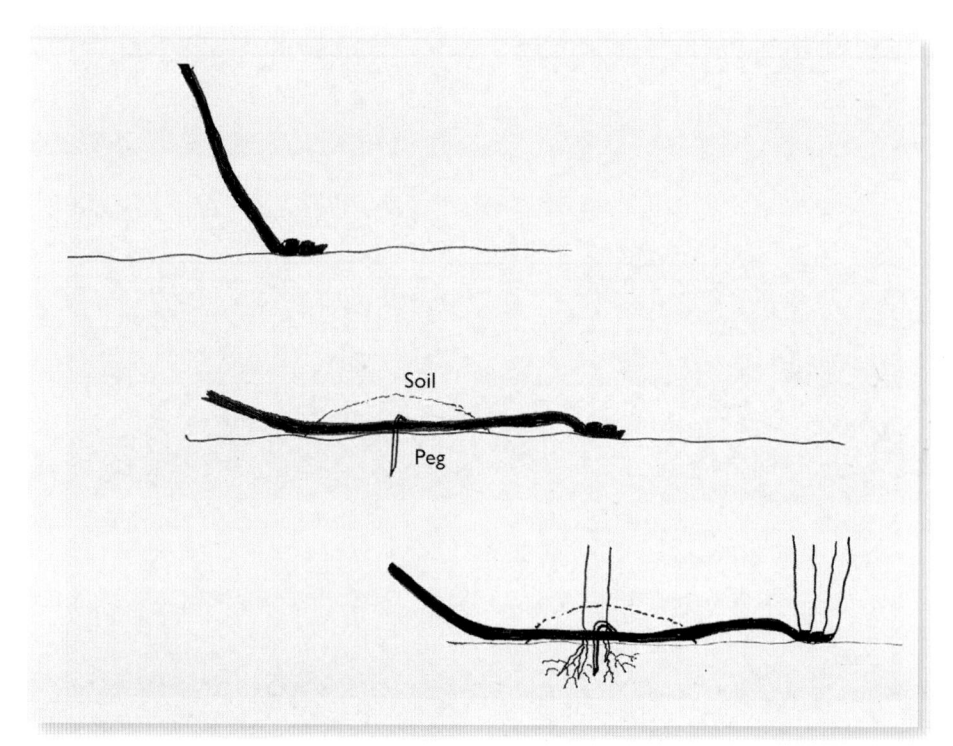

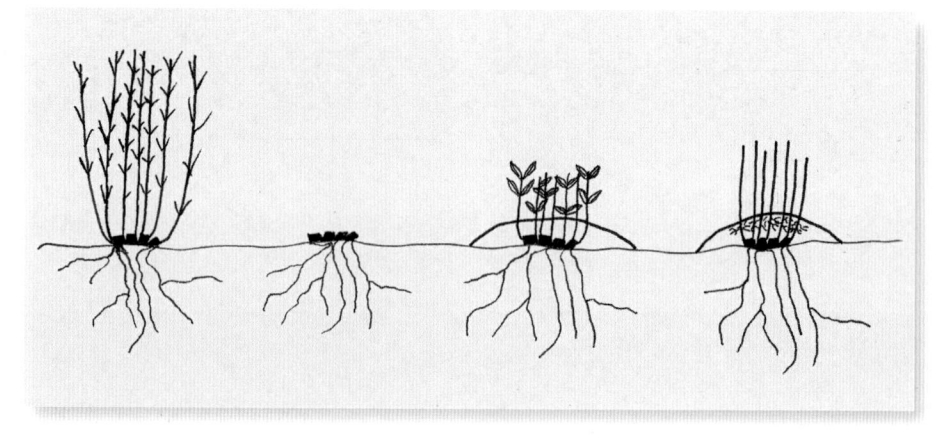

Stooling

Stooling is best achieved by covering over a stool with soil as the regrowth begins. The new shoots will then grow with roots formed in the soil. These can be cut and lifted and used to fill up gaps in the coppice. After they are established, the whole coppice can be cut again (hopefully taking a cut of a few sticks for walking sticks) and the old stools will draw up the new trees. Being maidens, these trees will make good standards in the longer-term.

Top:
Layering, carried out as in example one.

Below:
Stooling.

Opposite right:
Wind blow of coniferous plantation, Tighnabruaich, Scotland.

HIGH FOREST MANAGEMENT

The rotation length of broadleaf high forest can be up to about 150 years for a veneer quality oak or beech on a poor soil. On a fertile soil, oak could be mature for quality timber after 80 years. If the aim is to manage a wood for 80 to 150 years before harvesting, then good management of high forest to reach high quality timber must start at an early age.

Thinning

Thinning is carried out to ensure the final timber crop is of the best quality at the end of the rotation. Thinning involves removing poorly formed or diseased trees and removing a nurse crop at the appropriate time. This ensures a maximum benefit for the final crop and provides more growing space. In upland coniferous plantations, thinning is sometimes not carried out as the risk of wind blow is too high. In a mixed high forest, first you must be clear about which species of trees you are growing on to maturity and then you need to decide which specimens you will favour and which you will remove.

Thinning should be carried out as a gentle process over a number of years to avoid the risk of wind blow and to cause minimum disturbance to the ecological balance of the woodland. The first thinning is likely to be carried out when the trees are about 25-30ft (7.6-9.1m) tall and the second thinning may not occur for another 10 years after this. The third thinning may be enough to leave the desired final crop of about 32 trees per acre (80 per hectare) with oak and 40 per acre (100 per hectare) plus with beech, for example. Each thinning will produce an increasing economic return.

Soil, aspect, etc. will affect growth rates, hence the span of years. For detailed information on thinnings, heights, ages, yields, consult the Forestry Commission booklet, Forest Management Tables. The table overleaf shows the typical development of a conifer stand through the stages of thinning, ending in a clearfell. This is based on initial stocking rate of 810 trees per acre (2,000 per hectare). (This is shown for your benefit if you happen to come across a stand. It is not a strategy of management I would recommend, as I explain below.)

Type of tree	Approximate age (years) at first thinning	Average age (years) at final felling
Ash	16-25	70
Beech	27-39	120
Oak	26-38	125
Corsican pine	18-30	55
Douglas fir	17-24	50
European larch	17-34	60
Lodgepole pine	20-34	60
Norway spruce	20-30	60
Scots pine	20-35	60
Sitka spruce	18-30	50

Thinning	Height (m)	Age (years)	No. of trees/ha after thinning	Produce
1st	9-11	15-25	1,500	pulp
2nd	11-14	20-30	1,100	pulp, fencing
3rd	14-17	25-40	800	pulp, fencing, small saw logs
4th	17-21	35-50	550	fencing, small saw logs
5th	22-25	40-60	350	fencing, small saw logs
Final	25-30+	45-75	0	saw logs

I have been managing a larch plantation for about 8 years. When I took on its management it was about 30 years old and had never been thinned and the result was a lot of tall thin trees. Tall thin trees have no market with the sawmills but to me who builds houses from roundwood poles this was a really useful plantation. The trees having put on very little growth had growth rings very close together. From a construction point of view this is exactly what I need from a durable softwood like larch. Had the plantation been thinned, the poles would have not been so strong for construction. I have been taking out about 30 trees per year and converting the poles into buildings. Because larch is a deciduous conifer, this has allowed natural regeneration to occur and the plantation is evolving into a more diverse woodland.

Felling Of High Forest

Clear Felling

This process has a devastating effect on the landscape and ecology of the woodland. A clear fell exposes the soil to the risks of erosion, raises the water table and leaves the surface area wet and the majority of wildlife is evicted from its habitat. Extraction methods that follow usually lead to increased ecological damage with large machinery carrying heavy loads across wet ground. Sadly this is still the most common method of felling mature high forest.

Group Felling

This process involves felling an area of trees of usually less than 1 acre (0.4 hectare). In a large woodland, several areas may be felled at quite large distances apart in a year. The benefits to the landscape and wildlife are greatly improved compared to a clear fell as the surrounding woodland offers new habitat, and visually the effect is far less dramatic. The areas can be replanted or left naturally to regenerate, though protection from grazing is needed.

Selective Felling

Selective felling is the most sustainable way of managing high forest, but it is also the most expensive and requires the most skill. That said, the benefits to the woodland and its flora and fauna should make this the only viable option for high broadleaf forest management in the future. The overall aim is to keep a woodland of mixed species trees at different ages and be able to select and fell specific trees, having minimum impact on the ecology and visual impact of the woodland. In this system, young, medium and old age trees grow together. If the felling of selected trees is tied in with the use of appropriate technology, like mobile sawmills and horse extraction, the damage to the woodland is minimal for a high forest system. Selective forestry must be based on the amount cut from the woodland and never exceed the regrowth. It would be a creative use of funding to see funds which are currently available for the establishment of monoculture conifer plantations, channelled into the management of high forests through selective felling. Selective felling needs to create openings at least the size of the canopy of a mature tree to allow natural regeneration to occur. If the openings are smaller, then shade tolerant species, such as beech and hornbeam, may be the only viable option.

Conversion of an Even-aged Plantation into a More Diverse Woodland

Foresters in Britain are not trained to be familiar with converting plantations into diverse woodland. This is because most training has been directed towards managing even-aged, monocultural plantations towards a clearfell and replant regime.

To make a successful conversion, the process will involve breaking up the

Opposite left:
Larch plantation showing natural regeneration of birch, ash, oak and hazel.

Above:
Roundwood timber frame building constructed from the larch plantation.

Management Of Woodlands 81

Old tree crop thinned

Natural regeneration occurs, more of the old tree crop removed

Young trees established, all old tree crop removed

regular planting structure of the plantation, changing existing clearfelling patterns and creating wide rides and open spaces. Suppressed 'weed' species are allowed to naturally regenerate, as light is let into the woodland floor, increasing diversity and moving towards an uneven-aged, mixed woodland.

One technique which can be used in the conversion is the 'shelterwood' system. This is characterised by the clearance of mature stands of trees in usually two or more fellings and the establishment of the new stand in the shelter of mature trees left from the previous felling. New species can be planted if the natural regeneration is limited in diversity.

Another technique known as the Bradford Hutt plan – developed in Tavistock in Devon by the 6th Earl of Bradford and Phillip Hutt, his head forester – involves a compartment selection system. Here the management is controlled by land area, not by timber volume. The plan is based on a mature tree having a canopy of 20 x 20ft (6 x 6m) and reaching an estimated maturity in 54 years. The species being trialed in the plan are Douglas fir, Western red cedar and Western hemlock. Working with nine compartments, the aim is to end up with one tree in each compartment maturing at 54 years. In year one, all of compartment one is felled and replanted with nine seedlings. In year six, the same is done in compartment two, in year 12 the same in compartment three etc. Thinning is carried out as an adjacent plot is felled and replanted. After 54 years the cycle begins again. If we understand the average age in years for maturity of a particular species and the average canopy size at maturity, we can use the Bradford Hutt model as part of a conversion process. The felling operation for this plan can be seen as a small clearfell but within the vicinity there are different stages of growth; seedlings, young crop, saplings, older saplings, and large trees. The woodland as a whole remains the same when an area is felled.

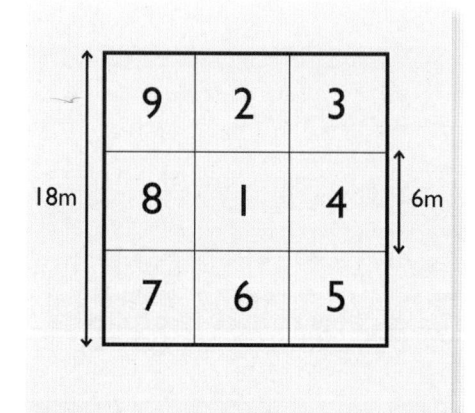

Top:
Shelterwood system.

Below:
The Bradford Hutt plan.

Another variation which has been tried in the tropics is the experiments of the Yanesha Indian co-operative in Peru, who manage an area of rainforest on a strip felling rotation. They have studied the forest and learnt that when a canopy tree falls it opens up a clearing which lets in enough light to allow natural regeneration to occur. They fell strips of forest as wide as a natural tree falling but of greater length, leaving adjacent strips as seed banks to maintain biodiversity. They are working on a 40 year cutting cycle (trees grow a lot faster in the tropics!).

Such a system could be incorporated into a plantation conversion. The strips could be felled following the contour and replanted with a mixture of species to add diversity.

Whichever methods are used to break up the regimented, monocultural, even-aged plantation, the long-term plan if timber is to be extracted should be a selective forestry operation. Income from selective forestry in terms of pure timber value, extraction costs, etc. will be lower than a conventional clearfell regime. Adding value to produce and a diversity of woodland activities will be needed to balance revenue losses; but the important economic projection will be long-term – there will be no loss of soil, species and fertility. In simple language, there will still be a woodland for future generations.

EXAMPLE

LECKMELM WOOD

Contact
Bernard Planterose
Leckelm Wood
Ullapool
Westeross
IV26 2TB
Scotland
www.northwoodsdesign.org

Size of Woodland
81.5 acres (33 hectares)

Type of Woodland
Mixed coniferous forest containing sitka spruce, Scots pine, small amounts of Douglas fir and lodgepole pine. Small gullies of native deciduous trees are scattered throughout the woodland. Planted in 1960 by the Forestry Commission, prior to that it was covered in oak, ash, birch, rowan, alder and willow.

Management Objectives
To promote and construct sustainable timber buildings that cope with the high rainfall of northwest Scotland, utilising roundwood columns where possible and to add value to the timber by constructing furniture. To reduce the amount of sitka spruce and plant more European larch as it doesn't need chemical treatment. To convert at least quarter of the woodland back to native hardwoods, predominantly ash, oak, cherry and alder. Management is based on a selective forestry regime with individual and small coupe felling. This will mean that an

Top:
A strip being harvested, Yanesha Co-operative, Peru.

Below:
Home and bothy for Bernard and helpers at Leckmelm woods.

extensive contoured ride system will need to be developed to enable ease of movement of individual trees.

What Has Been Achieved So Far

A challenge of working on the land was the time to progress projects whilst the children were young and needed much of their time. For a number of years the forest management fell on Bernard alone, but now with children grown up and able to operate chainsaws, there is potentially a team of five operators ready to contemplate the next phase of management.

A lot of resources have been spent on improving access for extraction (1:7 slope). Extraction with the Husqvarna Iron Horse has been superseded by an excavator with timber grab to enable larger scale harvesting operations to be carried out. Restocking no longer includes ash, since the outbreak of *Chalara* and species selection is favouring cherry and sycamore with wych elm and birch and possibly small leaved lime. Bernard is planting hardwood species in the hope that his business has a life beyond his own and these are timbers he wishes he had growing now for furniture and flooring.

Many buildings have been constructed locally and within the woodland itself. A workshop and office and small milling shed and yard form the main area of woodland activity. They have built a number of cabins and sheds with timber from the forest but lacking in larch has meant that they've always had to import some too. However, the drastically increasing price of larch and Douglas fir in Scotland has impacted on the way they design buildings and the way they are looking at the timber resource in the forest. The balance of the equation has shifted subtly and now they are beginning to mill their own spruce for cladding.

Top:
Bernard Planterose moving timber with the Husqvarna Iron Horse.

Below:
Micro-home design, one of the buildings from Northwood Design at Leckmelm Woods.

Not yet bold enough to use it on commercial projects, they have dressed and painted it on all four sides and made a very respectable cladding for their own private projects. This approach of making use of 'what we have got' is likely to become a pattern of future timber use.

Longer-term Objectives

They recently won an Scotland Rural Development Programme (SRDP) award for a new workshop for the business which will hugely assist their construction side and help develop and diversify manufacturing. They already design and build everything from bridges to houses but want to focus on some specific manufactured structures such as 'pods' (small self-contained accommodation/home work units), cabins/micro-houses and items such as bus stop shelters. The workshop will be 200m² and be heated by a biomass boiler burning logs and wood waste. They also plan to install a 6 metre kiln to dry their own wood which hopefully will enable them to use it in commercial projects as well as for furniture. Equipment will include a four-sided planer and large table saw as well as a 150mm circular resaw which will make milling on a Wood-mizer more economic. There is a Wood-mizer owned by another business resident in their yard which they can hire and, if they reserve it for initial milling only, they can then dry timber to resaw on demand with their own labour/machinery.

They have cleared another hectare of conifers for a forest garden and orchard. This will be terraced to deal with the slope and accommodate a couple of polytunnels. It will need fencing and the creation of ponds to irrigate the fruit trees, soft fruit bushes and polytunnels. It is a very long-term project but they aim to grow elderberry, blackberry,

Above:
***Early wooden studio,
Leckmelm Wood, Westeross.***

black and red currants, and to establish some top fruit, probably plum, pears and cherry in a large tunnel with netting sides and top cover. Resources only allow them to make small slow steps with this project, but one day it could become someone's livelihood, adding another dimension to their already richly diverse forest existence.

Leckmelm Wood is one of many Highland woodlands that suffered experimental Forestry Commission plantings in the 1960s. The Commission then decided to sell off the wood because with difficult access the cost of felling and extraction was higher than the value of the timber. It takes imaginative marketing of produce and, of course, adding value to turn these economics around. Leckmelm Wood is a pioneering example of what to do with a coniferous plantation. The example of selective forestry, appropriate technology, local marketing, adding diversity and dwelling sustainably within the forest gives us a model we can all begin to learn from.

Their cabins, long-term vision and timber buildings are a fine example of how to use the apparently 'valueless' resource and turn a problem into a solution.

Managing For Wildlife

Managing woodland for wildlife may involve active annual management, as in a coppice woodland, or minimum intervention, as in a naturally regenerating woodland and also in ancient woodlands. It may be specific management to protect the habitat of a particular bird or butterfly, animal or plant, or it could be to break up the regularity of a monoculture and vary the height and choice of species in the woodland to encourage greater diversity of wildlife in general.

Minimum Intervention

Many ancient woodlands have created wildlife habitats purely by nature's evolution and in these situations we should allow nature to continue her fine work. Areas of natural wilderness are few and far between in the British Isles. Where they exist we must leave them as such and also consider choosing new areas within other woodlands where we can 'fence' ourselves out.

Coppicing For Wildlife

Purely from its historical pattern of being a practice of woodland management for thousands of years, it means that coppice woodland has a unique variety of flora and fauna adapted to its cycles of cutting, including some species found only in working coppice woodland. For example, many of the fritillary butterflies feed on violets which appear only when the coppice is young and in cycle. Once it

Left:
Veteran oak, Prickly Nut Wood.

Below:
Scalloped edge adjacent to woodland ride.

Over page:
Woodland pond, Prickly Nut Wood.

becomes overstood the violets are shaded out and the food source is lost. Key points to consider when managing coppice for wildlife will be the size of the cants and the length of the cutting cycle. Having areas of coppice of different ages throughout the woodland will help to support a wide range of different needs. Leaving standards creates more diversity and dead standards are good for woodpeckers, nuthatches and may have holes for bats.

Dead Wood

Dead wood provides habitat for insects, fungi, mosses and lichens, invertebrates, amphibians and small mammals. Piles of dead wood are best left out of direct sunlight as the moisture content is an important part of the habitat for many species. Leaving dead wood in different areas, like on the edge of rides as well as under the canopy, will help encourage diversity of species. Leave standing dead wood where possible.

Veteran Trees

Veteran trees are defined as:

> *"Trees which, by virtue of their great age, size or condition, are of exceptional value culturally, in the landscape or for wildlife."*
> English Nature

These trees are rare examples which have escaped the forester's pattern of only seeing trees as timber and have been allowed to live out their lives to their full potential. If we see ourselves as part of the ecosystem (which we should!) as opposed to some superior species outside it, then, in order for the ecosystem to be healthy, all species should have members of different ages. Our own society would be unhealthy if we decided that maturity was fixed at say 50 years and beyond that people should not continue to live!

Veteran trees are landmarks. Some are well known like the Major Oak of Sherwood Forest, others are just known to a few of us who visit them as a place of contemplation and spiritual quest. Such trees also are host to many of our rarest species of lichens, bryophytes, fungi and invertebrates. In every woodland we should be making space for veteran trees to grow old with dignity (see the Veteran Tree Initiative in Useful Organisations in Appendix Eight to register a veteran with English Nature's database).

Ride Management

Rides and open glades within woodland often develop a unique flora and fauna very different from the rest of the woodland. Rides create corridors through the woodland where sunlight can penetrate and naturally species selection evolves differently. The woodland I work is predominantly sweet chestnut coppice and the ground flora is limited by the dense shade cast by the chestnut leaves, but the rides between the cants are covered with herbs and flowers and attract a wide range of butterflies (see Table of Butterfly Species in Chapter Three).

To enhance rides, a system of cutting an ascending profile into the woodland allows more light to the grassy paths, and creates a shrub layer between the ride and the main woodland. Scalloping can also be used to break down the linear effect of straight rides and reduce the wind tunnel effect that can sometimes occur. In coniferous woodlands widening rides should be done gradually, as often the only surviving broadleafs are growing on the edge of rides and may already be a vital habitat at their present size for certain species.

Water

Streams and ponds greatly improve species diversity in a woodland. Water in woodlands should not be overshadowed and the southern aspect should be open in many places to allow sunlight in. The banks should be gently sloped with shallow marginal areas leading into deeper waters. Ponds will encourage wildlife such as herons, ducks, occasional kingfishers, dragonflies and, of course, frogs and other amphibious creatures. Marginal plants like the attractive flowering rush (*Butomus umbellatus*) allow dragonfly larvae to climb out of the water, and the tubers are edible when cooked, containing a high concentration of starch. The true bulrush or common club-rush (*Schoenoplectus lacustris*) is a useful plant to grow in the margins of woodland ponds as it can be cut and used to weave seats for rustic chairs. It avoids the need to purchase sea grass imported from China and the Far East. Carp and crayfish could be considered in a woodland pond, if there is continuity of water and some sunlight, as they can be an added source of protein for the forest dweller.

Nesting Boxes

Nesting boxes can be added to encourage particular birds or bats into the woodland. I have positioned bat boxes near to my craft areas so that they will help keep the midges and mosquito numbers down during summer evenings! Rides through coniferous plantations are a good habitat for bats and an ideal place for positioning boxes. Bird boxes for owls, woodpeckers and kestrels can be constructed if there is a lack of dead or hollow trees and branches within the woodland.

FROM TREE TO FINISHED PRODUCT

Chapter Six

A hurdle maker who cuts seven year-old hazel in cycle needs nothing more than a billhook and a sharpening stone to manage a woodland and make the finished product. An investment of maybe £25 which will returned on the first hurdle.

But not all management is that simple, as some woodlands involve moving heavy stems of timber, dealing with leaning and hung up trees, awkward access and ecologically sensitive sites.

Cutting

Traditionally, all trees were cut with axes and hand saws prior to the introduction of the chainsaw. I cut large coppice stems with a chainsaw. Chainsaws are noisy, polluting and very destructive in the wrong hands; but used with discretion, they are a highly developed piece of technology and have a place amongst most woodsmen I know. I have met many keen coppicers who start their first season with a bow saw. Some stick with it, but the majority have a chainsaw the following year. I cut short rotation coppice with a billhook, up to about 2in (5cm) with a bow saw and beyond that with a chainsaw, but I do prefer cutting the shorter rotation by hand. Most modern chainsaws have catalytic converters and use biodegradable vegetable-based chain oil, and a 2-stroke synthetic low smoke oil. This is particularly important when coppicing and working near rivers.

A recent addition is the battery chainsaw. I have been using a Stihl MSA 160C to cut short rotation chestnut coppice. With two 36 volt lithium ion batteries that I can recharge from my solar system, I have been able to successfully cut three acres of coppice without the need to use petrol, and at a commercial pace. As battery technology further improves, the battery chainsaw is likely to be a welcome addition for most people working with small dimension timber. A two person crosscut saw is the alternative to a chainsaw for larger timber and if kept well doctored will keep two people very fit! Bow saws are ideal for 2-3in (5cm) coppice and the 21in (53cm) triangular shape is the most useful to get in amongst the rods.

The billhook is the perfect tool for cutting short rotation coppice or in cycle hazel. Make sure you find a real forged tool and not a mass produced modern replica. There is no comparison, as the metal on the modern replica will not have been tempered (heating and cooling of the metal during forging to achieve a specific hardness), and the key to success with all these tools is sharpness.

Milling

The milling of large timber was traditionally carried out by pit sawing (and still is in many parts of the world).

An old saw pit site still exists in the woods where I work and I was fortunate enough to meet one of the cutters who worked there. It was, without doubt, highly skilled and very physically demanding work. The arrival of the steam powered sawmill must have been a relief to many.

One of the dilemmas in forestry practice is the extraction of timber, as taking large logs out of the woodland does cause a lot of damage. Large machinery is needed and the woodland rides get compacted. One answer to this on sensitive sites is to bring the mill to the log. There is now a wide range of portable sawmills available, and many operators will hire themselves and their sawmill on a daily basis. The cost of this tends to be high compared with delivering to a static sawmill but this must be weighed up against the benefit to the woodland flora. The most basic mill is a chainsaw mill, a one or two person operated chainsaw and frame which can be carried with no more damage than footsteps to the most awkward of places. The mill works with a rip chain and is therefore quite wasteful unless you have a good market for sawdust, but it is very cheap compared to other alternatives.

For a high quality finish, the mill needs a band saw blade. Mobile bandsaws such as the Wood-mizer and the LumberMate are easily towed with a four wheel drive vehicle, set up in a woodland and they produce an excellent finish. Mobile bandsaws cost from £6,500-£25,000.

Another option is the Lucas Mill or Forester Swivel Saw, these are plate blade sawmills that can fit into an average pickup, be set up by one person and produce a good quality finish. It has the advantage over a bandsaw because the blades can be sharpened by the operator rather than by a saw doctor. (Some bandsaws now come with a self sharpening kit.)

These mills could be a worthwhile investment for adding value to woodland and could form the basis of a small-scale timber business. This could be one possible way of making a coniferous plantation viable while you are in the process of conversion to a more sustainable woodland.

Opposite left:
Nick (The Hurdle) Le Dieu, cleaving hazel with a billhook.

Top left:
Selection of billhooks.

Top right:
Pit sawing in Honduras.

Below:
Chainsaw mill in operation, Papua New Guinea.

From Tree To Finished Product 93

Extraction

Large harvesters are the combine harvesters of the woodland, as they leave their compaction marks wherever they have been. Mechanical extraction is hard on the environment. Some smaller machines such as the Iron Horse are one person operated and drive on caterpillar tracks that are less damaging than most. There is a range of small-scale Alpine tractors with timber trailers which cause minimal damage and are a good choice in many small woodlands. The most appropriate form of extraction is without doubt, however, still the horse. The extraction of timber by horses is known as snigging, and was the traditional extraction method in this country. With increasing awareness of the environmental impact of mechanical extraction, snigging is becoming a viable alternative in sensitive or awkward extraction sites. In Germany, timber extraction with horses is subsidised. In Britain, the horse logger has to compete with the mechanised alternatives, so having the most efficient equipment will improve daily productivity as well as putting less strain on the horse. Traditional snigging implements like using traces and a swingle tree are effective, but have been outdated by some of the Scandinavian designs.

The wheeled sledge, which is self-braking and has a 6ft7in (2m) bed with 4ft11in (1.5m) extensions, can be used as a two or four wheel sledge running on skidders if the wheels dig in. The braking system works on shafts. When going downhill forward skidders push the shafts into the ground. A cheaper option is the timber arch. It works as a gatherer and is more mobile than the wheeled sledge but it is harder work for the horse.

Using horses also supports the continuation of the breeding of

Top:
The author milling up a larch butt with a LumberMate 2000, Prickly Nut Wood.

Below:
Lucas Mill in operation at the Weald Wood Fair, East Sussex.

94 The Woodland Way

our working horses. Our oldest native breed, the Suffolk, is seriously endangered and the Clydesdale is rarely seen. The Working Horse Trust (see Useful Organisations in Appendix Eight) on the Sussex/Kent border is a charity set up to breed and train both horses and potential horse loggers in the art of snigging, and is a good starting point for gaining information, contacts and experience. The British Horse Loggers are a registered charity who set standards and offer training for new people coming into the industry. (see Useful Organisations in Appendix Eight).

In my own woods, I use a number of different methods of extraction, from carrying out timber 'on the shoulder', tractor and forestry winch, Alpine tractor with trailer and horse. Every wood is different and extraction methods will vary depending on sensitivity of the site, slope, and availability of equipment.

Many woodlands have been degraded by poor extraction techniques and extracting in poor weather conditions. So choose the right weather window and the right equipment appropriate to the woodland.

USING GREEN WOOD

When a living tree is felled, the wood is at that stage known as green wood. From that point onwards the wood begins to season (put simply, it loses moisture). Different species have different moisture contents and as the wood dries out it shrinks.

When wood is worked and shaped while it is still green, it still shrinks and distorts as it dries, but it is less likely to split than wood which has been cut into planks using a sawmill. Our education makes us think of starting with a plank,

Top:
Alpine tractor with timber trailer. The seat and steering wheel are on a turntable so you don't have to turn the equipment around when coming back out with a load.

Below:
Pete Albon with Baron extracting an oak butt, Castle Goring Estate, Sussex.

From Tree To Finished Product 95

cutting the plank to size and then shaping it to make our end result. When we use green wood, we can start with a tree or branch and the wood will be softer to work with hand tools, will cleave and bend more easily than seasoned wood. Cleaving is the process of splitting a log by forcing the fibres apart along its length.

The result is a far stronger piece of wood than one which has been sawn as the fibres have not been cut. Working green wood is a fulfilling experience starting with the raw materials of a tree and ending up possibly with a chair, table or other desired object. For information on getting started with working green wood, Mike Abbot's *Green Woodwork* is an excellent introduction.

Seasoning Wood

Seasoning or drying wood removes water, reducing the moisture content. Shrinkage in the wood takes place during this drying process. The weight of the wood is reduced and the strength of the wood increased.

For many of us, seasoning wood involves air drying. Where wood has been converted into planks it must be stacked carefully on firm level ground. A solid base or large bearers are needed. Bearers should be placed at a specific distance depending on the thickness and species of wood being stacked. But as a rough guide a 2in (5cm) thick plank should have bearers every 18in (46cm).

96   The Woodland Way

The sticks used to separate the planks should be placed directly above the bearers. The stack of planks should have air circulating above, under and through the stack and it should be protected from rain. Wood carefully stacked will air dry satisfactorily over a period of time (depending on the size of planks, species etc.). Air dried timber will hold some moisture content and whether air drying is enough or kiln drying is needed will depend on the end use of the timber.

Speeds of air drying can be increased by using solar drying techniques. David Blair's raised polytunnel at the Dunbeag Project (description of the project follows on the next page) uses solar drying and draws on good air circulation by positioning the polytunnel on a slope.

As the table shows on page 121 there is a lot of variation depending on the position of the timber which is air drying. I hope it at least gives some indication of woods which dry relatively quickly and those which don't.

The expensive option is kiln drying which for most of us means taking timber to a large woodmill. Kiln drying involves placing timber in a container with a control of air supply and air flow direction by fan and the use of dehumidifiers, and of course heat. The advantages of kiln drying are the control of the moisture content of the wood and speed of drying – days as opposed to months, or years.

If you are preparing wood to sell for building construction, then kiln drying

Opposite top:
Frankie Woodgate with Jeton and Yser unloading with the timber crane.

Opposite below:
Cleaving chestnut with a froe.

Above:
Air drying timber in a well-made stack.

Right:
Solar drying polytunnel at the Dunbaeg Project, Tighnabruaigh, Argyll.

From Tree To Finished Product   97

Approximate time for air drying hardwoods of 1in (2.5cm) thickness from green wood to 20 per cent moisture content

Species of hardwood	Time taken to dry (months)
Alder	3-9
Apple	6-15
Ash	4-10
Beech	4-10
Birch	4-10
Box	9-18
Chestnut (sweet)	6-15
Elm	4-10
Hawthorn	6-15
Holly	6-15
Hornbeam	6-15
Lime	4-10
Maple (field)	4-10
Oak	9-18
Pear	6-15
Sycamore	4-10
Walnut	5-12

will usually be necessary for timber which is going to be used for floors, doors, stairs and window frames. *Wood and How To Dry It*, a Fine Woodworking publication, has some interesting designs for home-made solar and dehumidifier kilns.

Neil Sutherland, an architect, has been converting a mixed coniferous plantation to saleable timber in Glenelg in Scotland. They plank the timber with a Wood-mizer mobile sawmill, have a purpose built drying shed, kilns and a workshop with a planer and moulder, and are able to sell added value products such as flooring direct to consumers. They are showing a positive alternative to the familiar cut and pulp policy and using Scottish timber to its fullest as can be clearly seen in this house they designed on their farm.

EXAMPLE

THE DUNBEAG PROJECT

Contact
David Blair, Michaela and Angus
Dunbeag
Tighnabruaigh
Argyll
DA21 2DU
Scotland
Tel: 01700 811 809
www.dunbeag.org.uk

Size of Woodland
30 acres (12 hectares)

Type of Woodland
Originally oak coppice with hazel and rowan underwood. The oak was coppiced for charcoal production to fuel the local gunpowder mill. Planted in 1963 with a mixture of conifers, mainly sitka

spruce, some European larch, Norway spruce, lodgepole pine, grand fir, Western hemlock and Corsican pine.

Management Objectives
To remove conifers amongst the oak coppice and add value to the conifer timber by mobile sawmilling and selling the timber planks. Pigs are being used to prepare the ground for natural regeneration after the felling of conifers. The woodland is to be used as an educational facility.

What Has Been Achieved So Far
Since 1995 areas of conifers have been felled and planked using a mobile Lucas Mill and extracted down the slopes by log chute. Much of the timber has been used to construct infrastructure buildings on the holding, the cabin, workshop, animal houses etc.

A raised polytunnel that grows some vegetables and acts as a solar dryer for timber has been erected. Stumps of spruce have been inoculated with the spawn of Indian oyster mushrooms. Pigs are used on site and are actively preparing the ground. Brash hedges are being constructed around natural regeneration areas. Fruit trees have been planted on a south facing clearing above the workshop and the garden is designed using permaculture principles. They are developing an agroforestry system to use

Opposite left:
The Sutherland's Scottish timber home, Glenelg, Kyle of Lochalsh.

Above:
David and Angus outside their self-built home at Dunbeag.

Above right:
Naturally regenerated rowan in the woods at Dunbeag.

Right:
Digging in a mile of pipe for the new hydro scheme at Dunbeag.

From Tree To Finished Product

as a model for future woodland crofts. Michaela is involved with Forest Schools and is working with local schools, engaging children with their local woodlands. Power is provided primarily by a newly installed 50kW hydro scheme which is also an important source of income.

The woodland is within a Woodland Grant Scheme (see Grants in Chapter Eight), to encourage the natural regeneration of the ancient oak forest that is a main objective in the management plan. The oak coppice is a beautiful remnant of the ancient forest of the peninsular and a climb through the forest to the rocky peaks above gives spectacular views out to Holy Island. Income at present comes primarily from the hydro scheme, the selling of plank timber and adding value through a range of local furniture and buildings, and the Woodland Grant Scheme.

Kilfinan Community Forest

A lot of David and Michaela's time has been working with a group in their local community helping to establish Kilfinan Community Forest. Tighnabruaich is a small community on the Argyle peninsula, surrounded by woodland, and has seen a steady population decline from 3,500 people to around 700 in fifty years. The Kilfinan Community Forest group see the woodland as an opportunity to encourage young people and families to live and work locally. Part of their plans are the establishment of six self-build forestry crofts to offer housing and work in the forest. So far they have fundraised and purchased 125 hectares of land from the Forestry Commission and have set up a community composting scheme and an allotment scheme. Tracks are being established to help with extraction and also to create footpaths and cycle routes through the forest. Their next project is to establish a timber processing yard

Top:
Primary school on a visit to Kilfinan Community Forest.

Below:
Community funday at Kilfinan Community Forest.

with an industrial grade milling machine and retail garden outlet, plus a new access road. Employment is being created and the forest is beginning to be managed and is helping to create a focal point of community stability in the rural area. They are a registered Scottish charity and are showing a pioneering example of a template for coniferous plantations in Scotland.
www.kilfinancommunityforest.co.uk

ADDING VALUE TO COPPICED WOOD

The majority of the coppice I work is sweet chestnut. The majority of coppice workers look at sweet chestnut as potentially fencing posts or split palings on a 12 to 15 year cycle and split rails on a longer cycle. When looking at a stand of chestnut, they are looking for quality straight poles which will cleave easily and of course for stocking rates. I make a few pales and I sell only a small quantity of fencing posts.

There are many people working chestnut for these uses and the price of the finished produce is low considering the labour input. If we look at the underwood sales from Maidstone chestnut coppice from 1987 to 1993 there is a clear decline in interest in buying the coppice and in the price of the coppice. The need to diversify and find new markets and new products from coppice woodland has never been clearer.

When I look at a stand of chestnut coppice, I see straight poles but I also look for natural curves and arches. If I cut a stand of 12 year old chestnut, the longest and straightest poles will go for rustic building or pergola construction. These are specific markets and usually I get the work of erecting the project, so I get a labour wage on top of the

Top:
Chestnut pales at Prickly Nut Wood.

Below:
Sweet Chestnut 'love seat'.

From Tree To Finished Product 101

year	total acres for sale	% sold	average price £/acre
1987	68.5	78	540
1988	69.8	54	872
1989	82.3	80	466
1990	90.3	51	450
1991	41.2	60	248
1992	46.9	67	128
1993	46.1	33	258

Source: Bioregional Development Group

price of the wood. Arches are stacked in separate piles and are made into rose arches or rustic furniture.

Curves are used for rustic furniture or for the sides of raised vegetable beds. Forks are saved of all sizes and used in garden seats or children's chairs or long ones for clothes props. Some large poles are cleaved and steam bent into the hoops for yurts and I look for walking sticks amongst the lop and top. Lop and top is also used in brash hedges, faggots or seasoned for kindling. Any wood not used after 18 months is converted to charcoal or occasionally used for firewood in the Rayburn.

To add to this, nuts are harvested and sold from standard chestnut trees, saps are tapped from many birches and maples growing amongst the coppice and I am presently looking at a medicinal market for barks and other by-products. By making a diverse range of products and establishing good local markets, I can return at least three times the value of 1 acre (0.4 hectare) of chestnut cut just for fencing posts or palings. By adding value wherever possible rather than selling the raw materials, it is feasible to earn a living from a far smaller area of woodland than is usually considered viable. When the woodsman becomes a

Looking back, it was time well spent. Having established a good customer base for one product, it is then easy to offer other products as they come on stream. Craft fairs and farmers' markets are always good places to sell value added products and you are selling direct to the customer and getting the full retail price for your product. Woodlands which have access off main roads are good places to sell direct to the public as well. For coppice workers, joining a local coppice group and having your products on their database can bring in orders as the groups often get large orders which they distribute to their members. The magazine *Woodlots* is good for advertising and selling standing, felled or planked timber as well as coppice products. If you plan to sell packaged products like charcoal, then investing in a well designed logo will help sell your products, both direct to customers and to the retailers. The internet has opened up huge marketing opportunities for those of us tucked away in the woods. A photograph of a product for sale on line can reach a huge market and keep us in the woods making products.

However you decide to sell wood or products, don't underestimate the time you need to put into marketing. If you work woods in one particular area and focus, as I do, on the local nature of your product – you will quickly build up a local market which will grow year by year. When marketing locally, you save on time and cost, and reduce your impact on the environment by reducing the transport of produce. I have turned down orders from shops because I

102 The Woodland Way

*Opposite below:
The author making chestnut faggots, Prickly Nut Wood.*

*Left:
Prickly Nut Wood, the name creates its own branding.*

Below: The author's rustic chestnut and hazel panels suit a cottage garden.

From Tree To Finished Product

is a winter activity. Although you may sell some of the underwood direct to a buyer, craftwork will add the value that makes it most worthwhile. Summer is the time of craft fairs and the opportunity to sell your crafts at a retail price and to meet other people, get new ideas and learn from other craft workers. Winter evenings and spring are therefore the time to make your crafts and to get products up and ready for the fairs. There are, of course, exceptions. Hurdle makers tend to cut and make as they go along (because hazel cleaves and twists better when it is freshly cut), charcoal burners get better results doing most of their burning from spring to autumn and selling throughout that time as the weather conditions are warmer and more stable and the market is at its peak.

Traditional crafts have a steady rate of sale, but creating new craft produce will give you added interest at any fair. Details of crafts and craftwork have been written up on many occasions and I am not going to cover more than a few basic ideas and useful tools and leave you to discover the rest. Finding balance is as always the key one searches for; the balance of time between coppicing and woodland management, craftwork and the marketing and selling of produce.

WOODFUEL

The interest and growth in wood as a source of fuel has grown dramatically over the past 10 years. The Renewable Heat Incentive is offering grants to those installing efficient boilers and through the Forestry Commission there is a woodfuel Woodland Improvement Grant (WIG) to put in new roads for timber extraction. All of this is driven by a need to meet targets in renewable energy.

In general this is a good development as it has to be a step in the right direction to be utilising our own renewable wood resource to provide heating as opposed to the world's dwindling supplies of fossil fuels; however when we look closer at the wood fuel picture I believe we need to be a little cautious and thorough in checking the source.

About two years ago, in my village in West Sussex, I was asked for directions by a driver of a lorry. I noticed it was a smart new Mercedes lorry, sign written with 'sustainable wood fuel' in bright lettering and was carrying wood pellets. So whilst giving directions I asked where he had come from and to my surprise he said, "Newcastle". I mentioned that seemed a long way to be bringing wood fuel considering he was delivering to the most wooded region of the country. His reply was even more disturbing, "This lot came over on a ship from Holland."

An example like this makes you question the sustainability of the enterprise; to take timber felled in Holland, convert the wood waste there into pellets, transport it across the sea, load up a large lorry, drive it from Newcastle to Sussex and back, to supply pellets for one boiler in the most wooded part of the country is at best ludicrous. This was, however, an early example and there are now local pellets made in the West Sussex area.

Another concern to me is the amount of boilers being installed in areas where there is minimal timber grown. If I take Cambridgeshire as an example with 3.5% tree cover, then a lot of installed woodfuel boilers will need fuel to be transported large distances across the country.

What is needed is a region by region approach. Areas with reasonable amounts of tree cover like Surrey (about 19% cover) are ideal for installing woodfuel systems. Those with lower tree cover should be focussing on a two stage approach. The first, to increase their tree cover and only then to implement the second stage of installing more wood boilers. The whole of the UK is well behind the European average of 37% with only 13% tree cover.

The woodfuel WIG grants from the Forestry Commission will be useful to large forest owners and estates where the new grant aided forest road system will help facilitate the extraction of timber and bring some woodlands back into management. I must point out, however, that these road systems are unsuitable for small woodlands.

So the interest in woodfuel is a good development, but let's make sure we manage it region by region. Let's not transport timber hundreds of miles burning up fossil fuels in the process which cancels out the benefit of using a renewable resource ... and above all let's plant more trees.

Charcoal Burning

I make charcoal as it utilises poor quality timber and is easy to sell. There is a growing market and it makes good use of all the wood that has not been used for other products, as well as offering an alternative to the large amounts of imported charcoal often coming from the Indonesian rainforests or mangrove swamps which is produced in poor social conditions. It is hard to justify coppicing a woodland purely for charcoal except perhaps if the woodland is primarily alder (as other markets for alder are hard to find and it makes excellent charcoal), or the extraction is very difficult. On-site conversion to charcoal and the extraction of the light weight product as opposed to heavy cord wood could make the coppice viable. Charcoal also makes use of overstood coppice material, in particular overstood hazel whose market when large and curved needs some imagination! Barbecue charcoal is the primary market, but artist's charcoal made from willow, spindle or occasionally oak is a useful high value product.

Charcoal dust is used as a soil sweetener by horticulturists and in

Opposite left:
Logs stacked ready for use.

Above:
Artist's charcoal.

From Tree To Finished Product 105

industry for filtration purposes. Biochar, the use of charcoal as an addition to the soil, is a growing market and a helpful addition to sell the large quantity of 'fines' (small size charcoal) that builds up in the wood when burning regularly. Blacksmiths will still buy charcoal for their forges if they make high quality items like hand forged tools. I use charcoal fines (sieved charcoal too small for barbecues) for pathways around vegetable plots. It creates a basking spot for lizards and is inhospitable to slugs which prefer not to cross it, as well as holding heat to radiate it at night to plants growing on the edge of the vegetable beds. For a substance used primarily for its value as heat, charcoal can also be used for cooling. I use a charcoal fridge which is a box surrounded by charcoal which I water at least twice a day during summer. The hotter the day the quicker the water evaporates, cooling the fridge box.

EXAMPLE

THE BIOREGIONAL DEVELOPMENT GROUP

The Bioregional Development Group (an environmental charity) has pioneered local resource-based projects in fuel, fibres and lavender production. The main interest for woodland has been the setting up of the Bioregional Charcoal Company.

Objectives

"To develop and promote locally produced sustainable charcoal for barbecues, as an alternative to imported charcoal produced in a way that caused deforestation."
From *Spotlight on Solutions – A People's Agenda*, WWF, 1997

The Bioregional Charcoal Company was set up in 1995 to link together scattered British charcoal burners and find major retail outlets for the charcoal. The Bioregional Charcoal Company was keen to support the traditional management of coppice woodland and saw charcoal as a valuable way of bringing overstood coppice back into cycle. Much of the imported charcoal at that time was also sourced as coming from ecologically fragile environments.

The Bioregional Charcoal Company were the first pioneering company to supply supermarkets and DIY centres with locally made charcoal direct from the charcoal burner who bags the charcoal and delivers direct to their assigned nearest store, therefore significantly reducing transport costs and pollution.

**Below:
Charcoal kilns at
Rawhaw Wood, Kettering.**

The first major break through was with the big DIY chain B&Q that had been dependent on centralised warehouses for deliveries to their stores, and this set a pattern for other companies to follow. The Bioregional Charcoal Company at its peak has involved over 70 producers across the United Kingdom and approximately 2,350 acres (950 hectares) of woodland are involved producing nearly a thousand tons of charcoal.

The Bioregional Charcoal Company estimates that each £20,000 of turnover creates one full-time woodland management job opportunity and with an estimated turnover of £700,000 in 1998, the equivalent of 35 full-time jobs were created. Some producers are wildlife trusts whose profits from the sale of charcoal go back into woodland conservation projects. The Cherry Orchard Centre for people with learning difficulties in Croydon, bags and delivers charcoal made from urban tree waste from London's first tree station which diverts tree surgery waste otherwise on its way to land fill sites.

A wood certification questionnaire is completed on all products which enables all wood to be traced to its origin. This allows for a certification scheme under the Forest Stewardship Council which includes woodland visits to ensure environmental quality of woodland management (see Certification later in this Chapter). Bioregional has now expanded its range of woodland products to include firewood and kindling materials, sold under its Bioregional Homegrown logo.

Contact
The Bioregional Charcoal Company Limited
BedZed Centre
24 Helios Road
Wallington
Surrey SM6 7BZ
www.bioregionalhomegrown.co.uk

Bioregional Charcoal Company Production 1995-1998

year	tonnes of charcoal (tonnes of wood)	Turnover £,000
1995	54 (324)	45
1996	139 (834)	122
1997	400 (2,400)	352
1998	800 (4,800)	700

Bodging
The pole lathe is once again becoming a more common site with woodworkers who need to turn some wood for a craft project. The quiet rhythm of the lathe is the perfect tool, easy to make and therapeutic to use. Along with the shaving horse, it forms the backbone of the woodland workshop. Most bodgers make chair legs and modern bodgers continue this practice whilst adding to a range of old and modern produce such as spoon blanks, baby rattles, candlesticks, rolling pins, baseball bats and complete chairs.

Steam Bending
Steam bending wood is a useful technique to create curves, hoops or any particular shape you require. The process involves ensuring that the wood absorbs the right amount of steam so that it can be bent to shape around a jig. A jig is made to the shape you need for a particular project. When steamed to the correct absorption level, the wood should bend without splitting and is then set to cool in the jig. The bending process readjusts the fibres which then reset in the new position. If you steam greenwood, the steaming process conditions the wood in a similar way to seasoning.

I don't know of anyone selling commercial steamers and most people I know make their own. My steamer is constructed from a piece of 4in (10cm) square box steel which is 15ft (4.57m) long, the perfect length for a 4ft (1.22m) yurt hoop. The steam comes through an army water boiler which boils 10 gallons (45 litres) of water at a time and the steam is transferred to the box steel by a laundry hosepipe, via a manifold. Both ends of the box open with large wingnut bolts, and a cardboard gasket can be used to improve the seal. The steamer is set at an angle with a drainage outlet at the lower end so as the steam condenses and turns to water, it can be drained off at intervals. A word of warning, my steamer has a pressure release valve of 50lb/sq in (3.53kg/sq cm). Remember that steam powers huge engines and at high pressure is potentially very dangerous – you have been warned. If uncertain, seek advice from a steam engineer. I have seen small steamers with a box more square in shape than mine and utilising a wallpaper striper to produce the steam. The design obviously depends on the length of wood you wish to bend. Large diameter thick plastic pipes can make a good steamer. With my steamer, I work on 1in (2.5cm) per hour for wood up to a 3in (7.5cm) diameter. However, this will vary on the output quality of the steamer. Experiment first with something that is not very important and learn about the habits and needs of your particular steamer.

Always take the wood out (wearing heat resistant gloves) while the steamer is at full capacity of steam and try to have the wood set in the jig within two minutes. Clamps and ratchet lorry straps are useful in helping this process. The best woods I have found for steaming are ash, sweet chestnut and yew.

WEAVING

I use the term 'weaving' loosely to cover all craft products which involve the weaving of wood. This includes the many uses of willow, from basketry to woven panels and living willow structures; hazel and its use in hurdles and woven furniture; oak for swill baskets; and lime or wych elm for the use of its inner bark (bast) for seat weaving, to name but a few.

JOINERY

Another broad range which includes rustic furniture, gate hurdles, Windsor chairs and just about anything else you can think of where pieces of wood are joined together. Always remember that green wood will shrink, so allow it to season in its finished state before constructing or, if mixing green and seasoned wood, use green for the mortised wood and seasoned for the tenon. With a slightly loose fit the green wood should shrink tight onto the tenon. Joinery is possibly the most effective way of adding value, but it can also be one of the most time consuming.

All of these craft categories should be symbiotic. For example, the bodged legs are jointed into the chair with a steam bent back and a woven seat and the off-cuts are converted to charcoal. A range of craft skills will ensure that you can gain the highest possible value from a small area of woodland.

Above:
The Roundwood Timber Framing Company's woodland steamer being used to steam bend roof rafters.

Opposite right:
The finished roof with steam bent rafters, The Sustainability Centre, Hampshire.

EXAMPLE

PRICKLY NUT WOOD

Contact
Ben Law
Prickly Nut Wood
Lodsworth
West Sussex
GU28 9DR
www.ben-law.co.uk
www.the-roundwood-timber-framing-co.ltd.uk

Size of Woodland
8 acres (3.2 hectares), plus management of adjacent 85 acres (34.2 hectares) under Forestry Commission English Woodland Grant Scheme Contract

Type of Woodland
Primarily sweet chestnut coppice, planted about 150 years ago with some oak standards and some mixed broadleaf woodland of oak, alder, birch, rowan and hazel. Some areas of derelict mixed coppice and a larch and Scots pine plantation. Prior to the planting of sweet chestnut coppice the woodland was oak standards over hazel underwood. (Other status: SSSI for mosses, lichens and ferns, dependent upon the coppicing cycle.) A new area of mixed broadleafs planted in 1992, including a wildlife corridor linking the woodland to Lodsworth Common.

Management Objectives
To maintain coppice cycles to enable the sustained production of marketable coppice, timber products and charcoal. To maintain and enhance the wildlife value of the woodland and to demonstrate that employment in the coppice industry is compatible with the conservation objectives for the site. To show that living within the woodland

110   The Woodland Way

Opposite page

Top:
Chestnut coppice at Prickly Nut Wood.

Lower left:
Wych elm seat woven from the bast of the wych elm, Moreton Wood, Hereford.

Lower centre:
Woven panel of chestnut lath, creating the formwork for earth plaster (wattle and daub) Prickly Nut Wood.

Lower right:
Oak Welsh stick chair by Stewart Whitehead, Llangollen.

This page

Right:
Lichens enjoying the environment of Prickly Nut Wood.

can be low impact and beneficial, both for the woodland environment as a whole and in terms of rural sustainability with the need not to have to commute to work. To develop local buildings from the woodland resource and to investigate markets for non-timber products like saps, fibres, nuts and honey and to market all produce from the woodland as locally as possible. To demonstrate and provide training in traditional and new coppice and woodland crafts. To control invasive rhododendron and to maintain the woodland rides to be of benefit for wildlife and butterflies. To ensure enough coppice is cut per annum to allow nesting sites for migratory nightjars.

What Has Been Achieved So Far

The woodland was acquired in 1992 through barter (an exchange of labour for land); and was at that time predominantly overstood with rampant rhododendron amid storm damage from 1987. One area had been cut for walking sticks and the stand was in good condition. Two larger areas of adjacent woodland are now managed with Prickly Nut Wood as one project. A Woodland Improvement Grant has been obtained to cut the last few acres of rhododendron, some of which was seasoned and then sold as kindling. There is now a minimal amount of rhododendron remaining with 50 acres (20 hectares) having being cleared. An old catchment pond which once fed a brickworks has been dug out and restored and adds a beautiful wildlife feature on the woodland edge. Wild ducks and an abundance of dragonflies are regular residents.

Clay from the pond was used to build a coppice wood forestry building whose self supporting roof structure is based on reciprocal hogan houses, and for the internal walls and fireplace of the Woodland House.

Nightjars are annual visitors and nest in the newly cut coppice. Tawny and little owls are abundant and grass snakes, slowworms and lizards are regularly seen basking during summer. The author has been resident in the woods since 1992, having lived in various benders and yurts. In April 1998, he was permitted a three year temporary planning permission to live in a caravan. On 4th June 2001, having satisfied the criteria in PPG7, he received full permission for a forestry dwelling, a cruck framed chestnut cabin with straw bale walls (see Relevant Forestry Planning Cases in Appendix Four for case reference).

Power from solar panels with some winter back up from wind turbines gives winter lighting, a hot shower, music and charges the telephone for taking orders and enables broadband connection. The forestry workshop runs table saws and a number of battery and other power tools, mainly during the summer months

Top:
Through the copse to the woodland house. This style of building – roundwood timber framing – has inspired many other buildings that are now part of the main produce and timber use from Prickly Nut Wood.

Below:
Forestry workshop, Prickly Nut Wood.

Birds recorded by the author at Prickly Nut Wood

Barn owl	Little Owl
Blackbird	Magpie
Black cap	Mallard
Blue tit	Mistle Thrush
Brambling	Moorhen
Bullfinch	Nightjar
Buzzard	Nightingale
Carrion crow	Nuthatch
Chaffinch	Pheasant
Chiffchaff	Pied wagtail
Coal tit	Redwing
Cuckoo	Robin
Dunnock	Rook
Goldfinch	Song thrush
Greater spotted woodpecker	Sparrow
Great tit	Starling
Greenfinch	Stonechat
Green woodpecker	Swallow
Heron	Tawny Owl
House martin	Tree creeper
Jay	Woodcock
Kingfisher	Wood pigeon
Lesser spotted woodpecker	Wood warbler
Long tailed tit	Wren

when there is a surplus of power.

Living in the woodland has helped control of deer damage and enabled a clear picture of the wildlife within the woodland to be built up. A compost toilet recycles all organic wastes and adds fertility back to the soil.

The woodland rides are rich with medicinal herbs and the first rows of coppice on each side of the rides have been cut on a short rotation for walking sticks to improve the woodland profile and allow extra light into the rides. Some chestnut standards have being singled and encouraged for nut production. Coppice fruit avenues mainly of apple varieties have been established between chestnut blocks to increase food production and increase future opportunities of sales of non-timber produce. An area of raised beds has established a productive garden to provide fresh produce all year round.

Charcoal production has been the backbone of the woodland business over the past twenty years, but now charcoal tends to be made from timber waste as opposed to cutting coppice to make charcoal.

Chestnut and larch are utilised in the projects of the Roundwood Timber Framing Company, (a specialist construction company set up by the author) which has built a number of houses from round poles in the local area. The setting up of the company has created a number of work opportunities for part time roundwood timber framers with up to six being employed by the company in the peak building season.

Extraction is carried out with an Alpine tractor and small timber grab and a LumberMate mobile sawmill converts some of the timber for building projects. Some of the chestnut is used for cleft post and rail fencing and lath for building works.

About 15 acres (6 hectares) approximately of derelict mixed coppice of hazel, ash, field maple have been restored having been derelict for 50 years and the oak canopy reduced to about 10% cover to create coppice with standards cants.

An apprentice scheme has been running for fifteen years and each year two people are selected to live and work at Prickly Nut Wood, gaining chainsaw qualifications and other woodland skills. A number are now managing woodlands in the local area.

A number of courses are run from the woodland each year ranging from woodland management through to roundwood timber framing.

Longer-term Objectives

To remove all traces of rhododendron. To manage the woodland with an open mind to the changes that may occur through climate change and plant pathogens. To continue to educate and promote the use of roundwood as a building material and to utilise local resources for local buildings. To carry out further surveys and build up more information on moth and butterfly numbers and needs within the woodland.

To leave the woodland diverse and ready for the needs of the next generation.

114 The Woodland Way

From Tree To Finished Product 115

Page 114

Top left:
Raised beds at Prickly Nut Wood.

Top right:
'Speckled Wood' being built for the National Trust by the Roundwood Timber Framing Company. (Sweet chestnut frame and shingles.)

Centre left:
Roundwood chestnut trellis from Prickly Nut Wood.

Centre right:
Cleft hazel and chestnut panels from Prickly Nut wood.

Bottom:
Sawn oak posts and lathe fence made on a LumberMate at Prickly Nut Wood.

Page 115

Top:
Bespoke Prickly Nut Wood gate.

Below:
Apprentices Barney Farrell and Tom Baker with chestnut yurt poles, sorted and ready for sale, Prickly Nut Wood.

This page:
Scribing on a Roundwood Timber Framing course at Prickly Nut Wood.

Opposite above:
Forest Stewardship Council Logo.

Opposite below:
Common lizard amongst the log piles, Prickly Nut Wood.

Page 118

Above:
'Withyfield Cottage' built by the Roundwood Timber Framing Company.

Marketing of High Forest Produce

I have emphasised the importance of adding value in relation to coppice produce. With high forest timber the knowledge of value will be the key in deciding what to do with any particular tree. Large timber is heavy and awkward to extract from the woodland and costly to transport to the sawmill, so on-site conversion should be the first option considered. But conversion into what? Quarter sawn planks? 2 x 2s? 4 x 2s? Or does the timber have another use with higher value? Identifying whether a tree is of veneer log quality (logs which are of straight, knot-free quality which are peeled into strips and then used to cover furniture of poorer quality wood) takes a trained eye and it may be a good investment to call in a local timber buyer to have a look at the quality before deciding what it should be used for.

So veneer quality logs and turnery burrs (a rough growth that develops on the side of the stem and can make beautiful bowls) may be your optimum market, and other sawlogs should be converted on site where possible. There are plenty of pulpmills that will buy truck loads of timber, but we need to move away from the sale of bulk loads of pulping timber which is the industrial forest model. The cost of transporting the timber is high enough in pure monetary terms let alone the environmental damage caused by the timber miles. The industrial forest model is based on the clearfell and replant regime and not on human-scale selective forestry with added value produce. If on-site conversion is not a possible option, then sale to the most local sawmill would be the next best option. For selling high forest produce, the magazine *Woodlots* deserves another mention.

Certification

The idea of timber certification came into being as a way to stop false claims and labeling of timber products. At the beginning of the 1990s there were many products on sale claiming to come from sustainably managed forests, but no mechanism to oversee whether the claims were valid or not. With organic food, certification is carried out by a number of certifying organisations like the Soil Association, and in time the public trusts the mark of the certifier on the products. In order to ensure the different certifying organisations are

From Tree To Finished Product

performing to a similar set of standards an umbrella organisation oversees the criteria and performance of the certifier. In the case of organic food this is the International Federation of Organic Agricultural Movements (IFOAM).

Certification for wood products has many similarities. Certifying organisations such as the Soil Association and SGS Forestry in Great Britain were some of the first organisations to certify the management of British woodlands. The Forest Stewardship Council (FSC) is the umbrella body that oversees these certifying organisations. One of the difficulties in certifying timber is that when a woodland is certified, the timber is often sold to many different places and made into many different products, so a chain of custody of certification is made to trace the timber from the woodland to the finished product.

The Forest Stewardship Council is an international non-profit making organisation that oversees the global certification of forests. The concept of certification and the economic incentive of the 'eco-stamp' on the product is now beginning to be realised. At the beginning of 1996, three forests in the UK were certified and a year later over 20 more were certified, with many more striving to gain certification. Now 45% of UK woodland is certified.

The size of woodlands varies from a large part of the Duchy of Cornwall's Estate to 32 acres (13 hectares) at Oaklands Park, managed by the Camphill Village Trust. One of the difficulties in the early stages of certification has been the cost of certifying small woodlands. The paperwork and visits necessary to certify a small woodland mean that costs are not dissimilar to those of larger woodlands and this has penalised the small woodland worker. Certifiers now have mechanisms in place, however, within their criteria to help small woodlands and where possible group certification of small woodlands enables the cost to be reduced as an inspector can cover more than one woodland in a day.

Since the launch of the FSC, many different methods of certifying have been considered, and the number of organisations can seem confusing. The Programme for the Endorsement of Forest Certification (PEFC), was launched in the late 1990s to help facilitate certification in Europe, and UK Woodland Assurance Standard (UKWAS) was launched in 1998. UKWAS is a standard approved by the FSC and PEFC to form the guidance for certification. The Forestry Commission will ensure new woodlands entering the English Woodland Grant Scheme (EWGS) meet the criteria of UKWAS.

Certification is a voluntary process through which woodland owners agree to inspection of their forest management practices by an independent certifying organisation to an agreed standard. UKWAS provides the standard in the UK to which the certifying organisation can measure compliance with sustainable forest management practices.

One model that we should consider developing in the UK is the bioregional certification model. In the area I live this could be a certifying scheme for The Weald, focusing on localisation as a vital part of sustainability. An example we can learn from is the Pacific Certification Council (PCC) in the bioregion of Cascadia in Canada/USA. The PCC is linked under the umbrella of the FSC and its certification criteria is for ecologically responsible forestry. It focuses on the certification of forests within the bioregion, building on its Cascadian identity and hence helping to look after the forests that sustain the region.

The success of the FSC and their certifiers can only be assessed by the long-term results of improved woodland management. It is early days yet, but this market-led mechanism may prove to be one of the most positive initiatives in sustainable woodland management for the future.

The two main challenges I foresee that it faces are keeping up the 'standards' for certification as pressure mounts as demand for certified products increases; and finding a mechanism to emphasise the more sustainable nature of a certified woodland whose produce is sold locally. Here I am looking at timber miles and although I see global certification as an important and necessary process in encouraging the sustainable management of the world's forests, a cupboard door that has travelled to a DIY store in Cardiff from a certified forest in South America should not be seen in the same 'sustainable light' as a cupboard door from a local certified Welsh woodland. Timber miles are an important consideration and any reduction in the travelling of a product as dense and awkward to transport as wood must be encouraged.

Sustainable Development in Woodlands

The majority of our timber (about 85 percent) is imported, yet around us lie derelict woodlands and unemployed people. It just doesn't add up! Imported timber comes from all over the world and unless the timber you are buying is certified, there is no way of knowing what ecological damage your purchase has done. It may be that the timber has been clearfelled from a poorly managed woodland. It may be affecting rare species or causing soil erosion. Socially it could be affecting the lives of indigenous people as your purchase could be stimulating illegal felling within indigenous reserves – common in many parts of the world. And whether it is certified or not, it is still travelling vast distances, using up fossil fuels before it eventually arrives in your locality. As with food, in order to live sustainably and provide for future generations buy local timber first and encourage local shops to stock local woodland products.

Hazel hurdles are reaching a new peak in popularity as a garden fencing product. These hurdles are made from coppiced hazel in the woodland. Buying them as opposed to a chemically treated softwood panel, continues to sustain an ecologically valuable habitat, provides rural employment, adds to the local look of the landscape and puts market pressure on landowners to plant less conifer plantations and more hazel. (Beware of cheap, imported hurdles in garden centres. These are generally identifiable by being fixed with nails and made from small diameter non-cleft hazel.)

Buildings can be constructed from our own woodlands. A revival in oak timber frame is certainly welcome, however the majority of oak timber framing companies in the UK import their timber from Europe. Alternative timber framing species such as Douglas fir and sweet chestnut should be encouraged as should be looking at alternative designs for construction, for example, utilising small diameter roundwood and coppice produce in the building industry.

The University of Surrey Civil Engineering Department has been involved in a European project looking at roundwood timber for construction. They have carried out strength testing for small diameter timber, structural systems, connections between roundwood and demonstration buildings.

Currently their work is with softwood, but sweet chestnut coppice grows as fast and is far more durable, and we have thousands of hectares of worked sweet chestnut coppice available. Chestnut has received a British Standard within the building industry enabling it to be used more often as it will fit in with Building Regulations criteria (reference Digest 445, *Advances In Timber Grading*, see Bibliography in Appendix Seven). As we start to use more of our local material so our cultural heritage is preserved and revived. The charity Common Ground promotes the importance of our locality in common culture and has pioneered a local distinctiveness project which seeks alternatives for the spread of uniformity throughout Britain. In the future, chestnut buildings are likely to be found predominantly in the south east of England as this is where the majority of chestnut coppice grows.

Meanwhile in north west Scotland, we can expect to see more like the Planterose's locally grown roundwood column buildings. These could become future distinctions of locality, as flint and local stone have been in the past.

FOOD FROM THE WOODS

Chapter Seven

Our earliest agriculture was as forest dwellers. We cleared the wildwood and began to cultivate the land. Forest dwellers (where they still exist today) use non-timber products (as well as timber products) for their needs – in particular woodland food – for the woodland has always been an integrated forestry agriculture system. Today woodlands are the source of some of our finest foods and foraging can bring delicious rewards. To me, however, woodland can also be designed to produce more than the wild forage food, and dispersed amongst the timber can be fruits, nuts, saps, fungi and meat, all part of a well designed woodland.

The natural woodland has a series of layers which together make up the woodland structure.

If planting a new woodland, creating a 'forest garden' (an integrated design of edible species using the woodland model) should be considered as part of the overall design. A forest garden can be incorporated into an existing woodland and form part of a rolling permaculture project whereby the woodland is slowly transformed over a number of years from a pure timber woodland to a mixed fruit and timber woodland.

Many food plants are naturally available within woodlands and others can be added as standards or in clearings or by making use of the ever abundant woodland edge.

I am not advocating that we should convert every woodland into a food and timber woodland, as many woodlands are a unique habitat and the introduction of species could upset the balance. But with plantations, both coniferous and broadleaf, and any monoculture woodland, the addition of fruit and nut trees should be seen as a benefit. Always think long-term when planting fruit and nut trees in a woodland as trees nearby may at sometime need to felled.

The following tables look at the different layers of the woodland and at native woodland food sources, as well as some possible useful non-native introductions.

Sweet Chestnut

The change in weather patterns over the last few years has given us warmer Septembers. This has improved the size and ripening of sweet chestnut although it still only performs well in the south of England. Coppiced chestnut will only produce a small crop once it gets to about 10 years of age, whilst standards on a good year produce a constant harvest throughout October and into November. I am so encouraged by the improving yield of sweet chestnut trees that I am leaving standards in every cant I cut. I sell chestnuts to our local greengrocer who will take as many as I can supply. Roasted chestnuts are a truly delicious experience and when carefully cleaned I also enjoy them raw. During October, hardly a day passes when they are not on the menu and therefore I have tried many recipes including pies, soups, nut roasts, bread and chestnut bread pudding and even wine. They can be dried to store or buried in the ground whilst still in their prickly nut cases and retrieved in the following spring. They will keep well in a basket if regularly turned. Honey from the chestnut flowers is creamy and delicious. There are many grafted varieties of chestnut that will improve the yield, often one large nut per case as opposed to the common three are produced. 'Marron de Lyon' is one well known variety. 'Marigoule' is a useful variety as it is partly self-fertile. These could be considered in a new planting, but do usually cost over ten times the price of a forestry whip.

Chestnuts have been a staple for many countries across Europe and into Asia, see analysis below.

Analysis of chestnut flour taken from Kew Bulletin No.44 in August 1890 by a Professor Church

Moisture	14.0
Oil and fat	2.0
Proteins	8.5
Starch	29.2
Dextrin and soluble starch	22.9
Sugar	17.5
Ash	2.6
Cellulose	3.3
	100%

Opposite left:
Highlighting different layers in the forest. Yarner Wood, Devon.

Above:
The late Robert Hart, pioneer of temperate forest gardening.

Food From The Woods

The Canopy Layer

Common Name	Latin Name	Edible Part	Comments
Ash	*Fraxinus excelsior*	seed	keys used to make pickle
Beech	*Fagus sylvatica*	nut/leaves	edible nut, presses into oil, leaves good in salads when young
Field maple	*Acer campestre*	sap	wine (slow to collect)
Lime	*Tilia* sp.	leaves/flowers	leaves delicious in salads, flowers make tea and fragrant honey
Mulberry ++	*Morus nigra*	fruit	eat raw, jams, pies, wines
Oak	*Quercus robur* *Quercus petaea* *Quercus ilex*	acorn/leaves	acorns ground for flour and roasted as coffee substitute (most palatable acorns from *Q. ilex*), wine from leaves
Rowan	*Sorbus aucuparia*	fruit	makes a sharp jelly and excellent wine
Silver birch	*Betula alba*	sap	excellent wine
Sweet chestnut	*Castanea sativa*	nut/flowers	edible nut, good honey crop
Sycamore	*Acer pseudoplatanus*	sap	wine
Walnut	*Juglans regia*	nut/leaves	edible nut, oil, pickle, wine from leaves
Wild cherry	*Prunus avium*	fruit	edible fruit (birds get majority!)
Wild pear	*Pyrus communis*	fruit	best cooked
Wild service tree	*Sorbus torminalis*	fruit	fruit makes wine and is edible when bletted
Yew	*Taxus baccata*	fruit*	sweet fruit* (the seed within the fruit is extremely poisonous)

\+ Acorn flour can be used in a similar way to corn flour and is best mixed with another flour or oatmeal. Holm oak seems to produce the most edible of acorns in this country as they taste less 'tanniny'. To help remove tannin, boil acorns in water before grinding. The bitter tannins are soluble and the water helps remove them.

++ I have had excellent results growing the hybrid red mulberry, *Morus alba x rubra* 'Illinois Everbearing', it has proved to be extremely vigorous and has produced fruit within two years of planting.

Top:
Sweet chestnuts opening.

Left:
Walnuts overhanging a footpath, Lodsworth, West Sussex.

Below:
Chestnuts cooking at Prickly Nut Wood.

Food From The Woods

Oak Leaf or Walnut Leaf Wine

This is an easy wine to make!

1 gallon (4.5 litres) of young oak or walnut leaves
1 gallon (4.5 litres) of water
2lbs (900 grams) sugar or 1lb 8oz (680 grams) honey
Yeast
2 oranges, 1 lemon

Pick the leaves when they are freshly opened and pour boiling water over them, leave for 24 hours and strain the liquid from the leaves, which should now be quite beautiful to look at. Boil the liquid and dissolve the sugar or honey, add juice of oranges and the lemon and a little grated rind. Start the yeast by adding the dried yeast to some lukewarm, sweetened water and leave in a warm place until active (half an hour usually). When the liquid is cooled to lukewarm add the sweetened yeast liquid and leave in a fermentation bin for one week. Then strain into a demijohn, seal with an air lock and leave to ferment. Rack once when sediment settles, bottle when fermentation is complete. It should be ready for drinking for the autumn equinox.

Above: Wine fermenting, Prickly Nut Wood.

Walnut

Juglans regia is a traditional timber tree in the British Isles as well as producing large crops of nuts, but not until it is well established. Grafted varieties such as 'Franquette', 'Fernette' and 'Ferjean' will produce far earlier and there are also walnuts known as heartnuts (*Juglans aliantifolia* var. *cordiformis*) which originate in Japan and are more frost resistant and may do better in the north of the country.

At Prickly Nut Wood I have experimented with growing the grafted walnuts 'Broadview' and 'Buccaneer' which cropped three years after planting, despite the slightly acidic soil. These have done reasonably well but the above mentioned varieties are more recommended. Grafted walnuts are expensive to buy, so once the tree is established it would be well worth taking cuttings for grafting (see Grafting on page 132). This will give you additional trees to plant and sell. Martin Crawford at the Agroforestry Research Trust has been growing a number of varieties of suitable nut trees for our changing climate and can recommend and supply suitable cultivars for your soil type and situation.

Hazel

Cobnuts are the more productive cultivars of the common hazel, *Corylus avellana*, and filberts are the productive cultivars of *Corylus maxima*; but unless grey squirrel control is already a key management strategy in the woodland the chance of getting crops of nuts is low. I tend to pick and eat hazel nuts before they have hardened, like the squirrels do, and eat them when they are moist and wet.

Traditional cobnut cultivation involves pruning to a goblet shaped structure and 'brutting' the new growth

The Lower Tree Layer

Common Name	Latin Name	Edible Part	Comments
Blackthorn	*Prunus spinosa*	fruit	wine and flavouring gin
Crab apple	*Malus sylvestris*	fruit	wine, jam, verjuice
Elder	*Sambucus nigra*	fruit/flowers	fruit makes jam and excellent red wine, flowers eaten as fritters, made into tea, wine or a non-alcoholic champagne
Guelder rose	*Viburnum opulus*	fruit	jam
Hazel	*Corylus avellana*	fruit/leaves	edible nut, salad leaves
Hawthorn	*Crataegus arnoldiana*	fruit	large sweet fruit
	Crataegus laevigata	fruit/leaves	leaves have nutty flavour when picked young
	Crataegus monogyna	fruit/leaves	fruit makes good addition to jams and a pleasant rose wine
Juneberry	*Amelanchier intermedia*	fruit	edible raw, cooked or dried
Juniper	*Juniperus communis*	fruit	jam and flavouring spirits
Medlar	*Mespilus germanica*	fruit	edible when bletted or baked or for wine

Above: Medlar, variety 'Nottingham', a reliable cropper at Prickly Nut Wood.

– this means partially breaking the branches downwards. This stresses the tree and encourages nut cluster formation.

In order to achieve yields of fruit from any species from the shrub layer, the need for light is an essential factor. These species will produce best at the woodland edge and in woodland clearings, on the edge of rides and where felling has taken place or an area of coppice cut. It may come as a surprise to know that gooseberries, redcurrants and raspberries are well naturalised woodland plants.

**Above:
Author harvesting cobnuts in Lodsworth, West Sussex.**

Food From The Woods   127

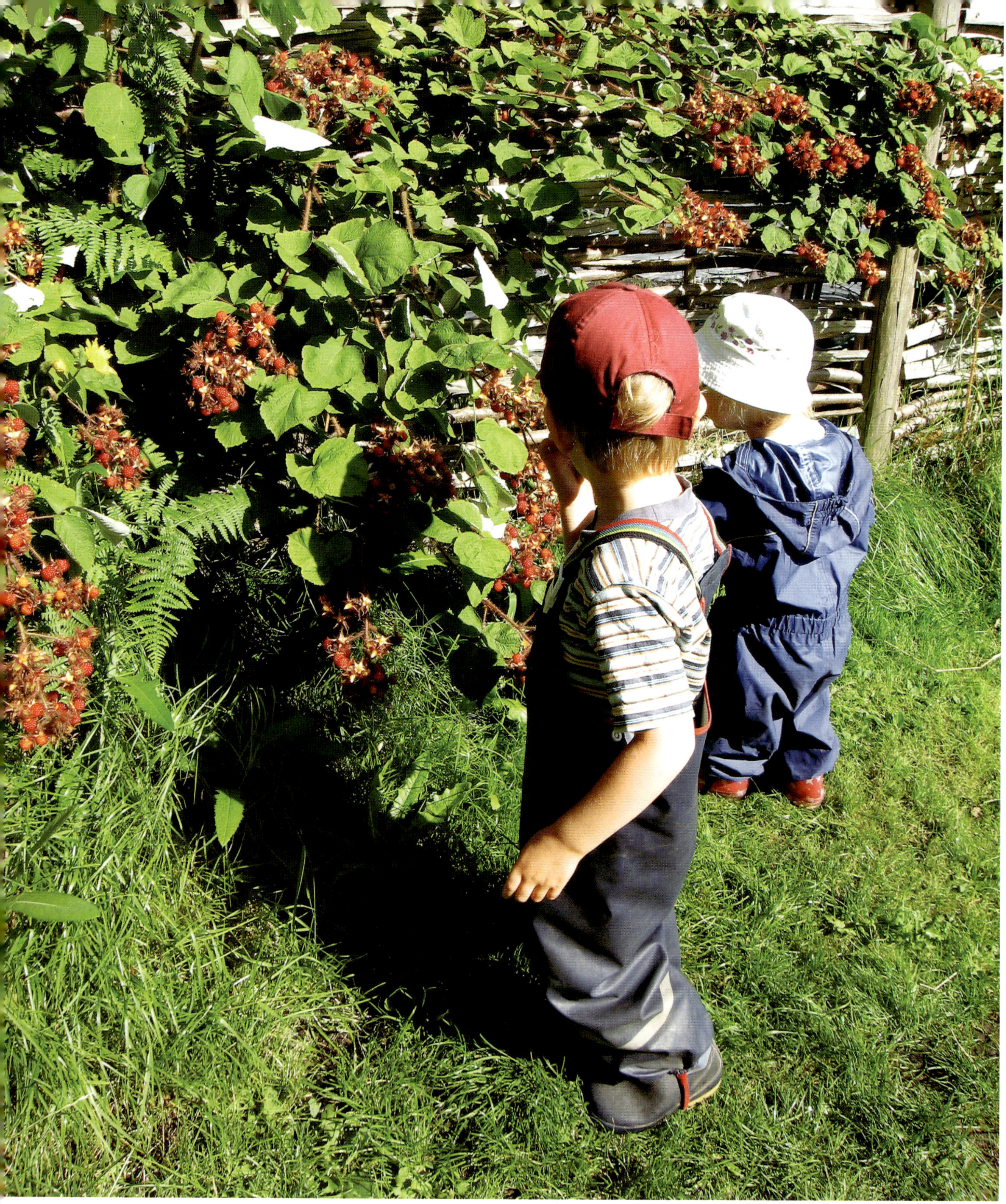

The Shrub Layer

Common Name	Latin Name	Edible Part	Comments
Bilberry	*Vaccinium myrtillus*	fruit	raw, cooked or wine
Currants	*Ribes sp.*	fruit	raw, cooked or wine
Eleagnus	*Eleagnus x ebbingei*	fruit	early edible fruit
Gooseberries	*Ribes uva-crispa*	fruit	raw, cooked or wine
Gorse	*Ulex europaeus*	flowers	wine
Oregon grape	*Mahonia aquifolium*	fruit	jam
Raspberries	*Rubus idaeus*	fruit/leaves	edible fruit, tea from leaves
Rose	*Rosa canina*	fruit/flowers	edible hips, petals
	Rosa rugosa	fruit/flowers	large edible hips/flowers

Opposite left:
Japanese wineberries, irresistible to children.

Right:
'Laxton's Superb' yielding well on MM106 rootstock in a coppice fruit avenue.

The Herbaceous Layer

Common Name	Latin Name	Edible Part	Comments
Alexanders	*Smyrnium olustratum*	stems/leaves	cook like asparagus or eat in salad
Burdock	*Arctium minus*	stems, roots	wine
Jack-by-the-hedge	*Alliaria petiolata*	leaves	salads, stir fries
Lemon Balm	*Melissa officinalis*	leaves	wine, tea
Meadowsweet	*Filipendula ulmaria*	flowers	wine, tea
Mint	*Mentha sp.*	leaves	teas, salads
Nettles	*Urtica dioica*	leaves	steamed vegetable, soup or beer
Oregon grape	*Mahonia aquifolium*	fruit	peeled young stems eaten in salads or as asparagus, beer can be made from roots
Pignut	*Conopodium majus*	tubers	edible tubers

A seasonal salad picked in Prickly Nut Wood in May consisted of:

- Catsear
- Chickweed
- Chives
- Fennel
- French sorrel
- Hawthorn leaves
- Hazel leaves
- Jack-by-the-hedge (garlic mustard)
- Lime leaves
- Marjoram
- Mint
- Oregano
- Ramsoms – flowers
- Raspberry leaves
- Rocket
- Wild sorrel
- Wood sorrel
- Yellow archangel

Ground Cover

Ground cover in woodland edge, rides and clearings

Common Name	Latin Name	Edible Part	Comments
Catsear	*Hypochaeris radicata*	leaves	salads
Chickweed	*Stellaria media*	leaves	salads
Dandelion	*Taraxacum officinale*	leaves, roots	salads, roots as coffee substitute
Golden saxifrage+	*Chrysosplenium oppostifolium*	leaves	strong flavour best mixed with other greens
Ground ivy	*Glechoma hederacea*	leaves	tea and ale flavour
Heather	*Calluna vulgaris*	flowers	beer
Pink purslane	*Montia siberica*	leaves	salads, cooked
Ramsons+	*Allium ursinum*	leaves, flowers, bulb	garlic flavoured leaves and flowers for salads and cooking, edible bulb
Salad burnet	*Sanguisorba minor*	leaves	salads
Sorrel	*Rumex acetosa*	leaves	salads, stir fries
Wild strawberry	*Fragaria vesca*	fruit	delicious!
Woodruff	*Galium odoratum*	flowers	steeping in wines, and laying in linen
Wood sorrel+	*Oxalis acetosella*	leaves	salads
Yellow Archangel	*Galeobdolon luteum*	leaves	salads, stir fries

+ denotes that plants do best in full shade

Left: Soft fruit from Prickly Nut Wood
Right: Wild strawberry, the real flavour of strawberry.

Food From The Woods

Climbing Plants

For climbing up trees and brash hedges, supports can be used for valuable species

Common Name	Latin Name	Edible Part	Comments
Blackberries	*Rubus fruticosus*	fruit	abundance of berries raw, jams, wines
Hop*	*Humulus lupulus*	flowers/leaves	female flowers used to make beer, leaves and shoots stir fried
Hybrid berries	*Rubus x ssp.*	fruit	same as blackberries
Japanese Wineberry	*Rubus phoenicolasius*	fruit	sweet edible berries, wine
Kiwi fruit* (hardy)	*Actinidia arguta*	fruit	eat raw

* can smother trees, good for climbing up dead trees

COPPICE FRUIT AVENUES

Coppice fruit avenues are where fruit trees are consciously planted between the cants of coppice, or in clearings next to cants of coppice. When a cant is cut, the sunlight reaches the fruit trees, causing the formation of fruit bud and a good crop should be produced the following year. It is possible to design the management plan of coppicing around the fruit trees, so that cants to the south of the fruit trees are cut more regularly, on a shorter rotation. This causes less shade to be cast on the fruit trees, and the cropping becomes more regular and abundant. A shorter rotation coppice is easier to fell without damaging the fruit trees between the cants. I have successfully grown apple trees on MM106 rootstock with some trees as close as 10ft (3m) from sweet chestnut coppice stools of over 100 years of age and achieved results in yield comparable to a commercial orchard! No sprays are used and one spade full of well broken down compost from the compost toilet per annum per tree is the only addition. The key to success is coppicing on a short rotation to the south of the fruit trees and ensuring careful felling of coppice at all times.

Grafting

Crab apples and wildling (pip grown) apples can be used as rootstock to graft on the 'scions' of a chosen domestic apple. This practice can considerably increase yield in woodlands and can also apply to grafting onto other rootstocks such as wild pear, wild plum, wild cherry and hawthorn.

The most useful types of grafting to use are whip and tongue for grafting onto young rootstocks, shield budding for summer grafts and cleft grafting (particularly suitable for grafting onto established wild woodland trees).

Whip & Tongue Grafting

This works best where the rootstock and scion (the cutting to be grafted onto the rootstock) are of similar diameter. The process involves making a 'tongue' on both the scion and rootstock so they fit tightly together and the cambiums are in contact. Seal the graft with wax or a tree pruning compound such as Arbrex and bind with raffia. This is best carried out when the sap is moving in March and April.

Left:
Harvest of 'Egremont Russet' from Prickly Nut Wood.

Above:
A selection of Prickly Nut Wood apples, picked and ready for sale.

Shield Budding

This involves inserting a bud from the present season's growth into a young rootstock. The bud is removed at high summer (end of June-August), inserted into a T-cut in the rootstock at least 6in (15cm) above ground level and the graft is tied with raffia.

Cleft Grafting (top-working)

Cleft grafting is best carried out towards the end of February and is ideal for converting established wild trees such as crab apple to a domestic variety. Cleft grafting involves opening a split across the diameter of a stem of the existing tree and inserting a pointed scion or scions into the wound. The wound should be plugged with clay and sealed with wax.

Oblique cleft grafting is the same process but avoids a split across the whole diameter of the established tree, and is therefore a less severe approach. With cleft or oblique cleft grafting the process involves grafting a number of scions onto the branches rather than a single graft with a young tree. This allows the opportunity to create family trees with, for example, different domestic apple varieties on the same crab apple root stock.

Planting

As trees are harvested from the woodland, some replacing with fruit and nut producing trees could be considered. Although the English walnut is slow to yield, there are now many grafted varieties that can start producing after about three to five years. For long-term thinking, planting some monkey puzzle trees as standards will ensure huge yields of nuts for the next generation. If planting top fruit such as apples, pears, plums etc., planting distances will depend on the rootstock used and in general distances will be much further apart than timber trees (see table below). With timber trees you plant close to get tall straight stems with minimal branching. With fruit trees you create space to allow the tree to branch as that is where the fruit will form. When planting a new woodland, it would be worth considering creating an area of forest garden. The forest garden is based upon the natural pattern of layers that grow in a woodland and puts together an integrated design of edible species using the woodland model.

BRASH HEDGES

I have mentioned the building of brash hedges to protect coppice regrowth. These hedges create vertical growing

Opposite right

Illustration 1:
Whip and tongue grafting.

Illustration 2:
Shield budding.

Illustration 3:
Cleft grafting.

Illustration 4:
Oblique cleft grafting.

Common Rootstocks

Common Apple Rootstocks

M25	vigorous	plant 20ft apart (good for cider apples)
MM106	semi-dwarfing	plant 15ft apart
M26	dwarfing	plant 8-10ft apart

Common Pear Rootstocks

Pear	very vigorous	plant 20ft apart (tall upright growth, good for Perry)
Quince A	vigorous	plant 12-15ft apart
Quince C	semi-vigorous	plant 8-10ft apart

Common Plum Rootstocks

Myrobalan B	vigorous	plant 15ft apart
Brompton	vigorous	Plant 15ft apart
St. Julien	semi-vigorous	plant 10-12ft apart
Pixy	dwarfing	plant 8-10ft apart

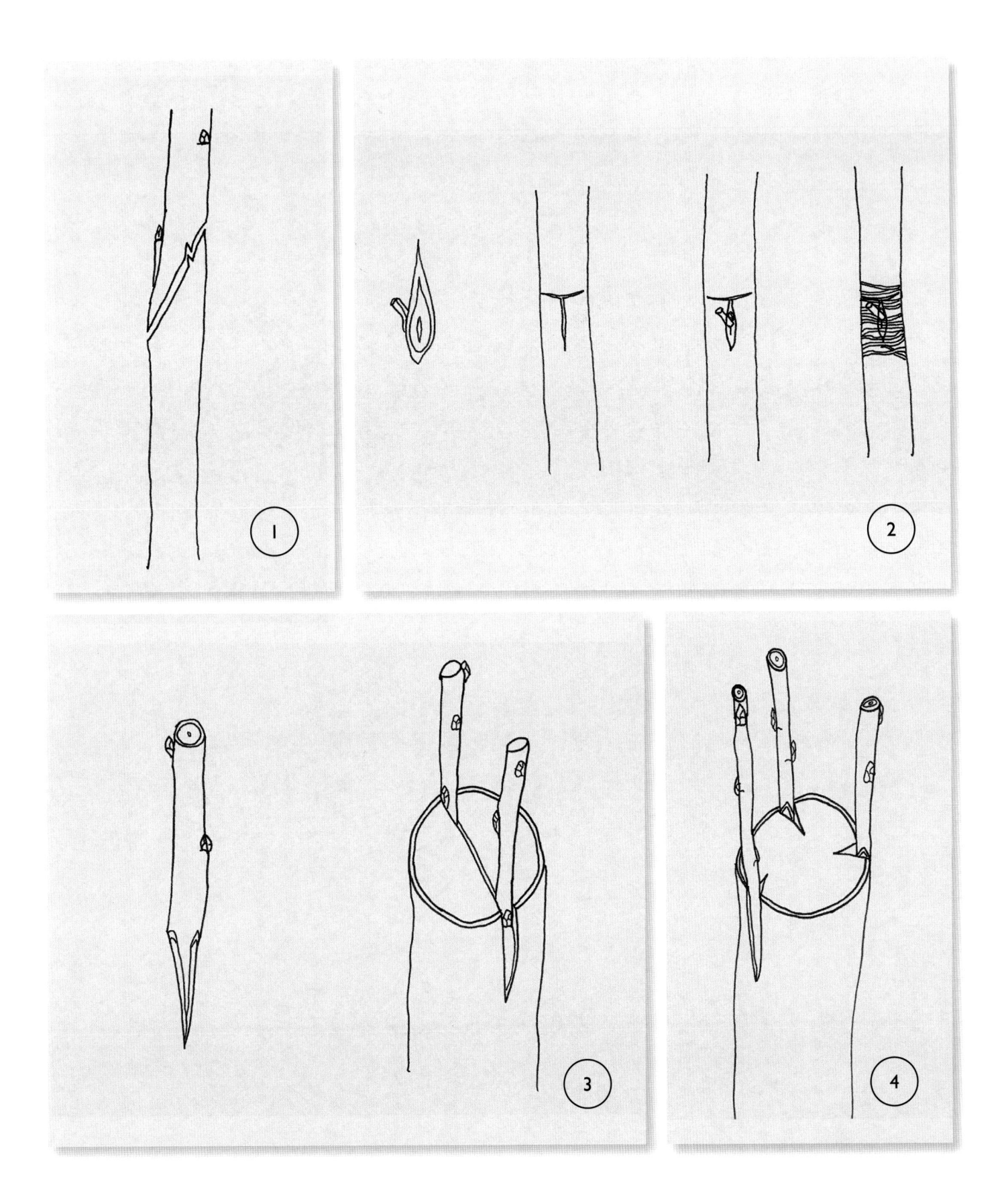

Food From The Woods 135

136 The Woodland Way

space and are often in full sunlight after the coppice has been cut. I transplant brambles to grow up mine as the brambles add to the protective nature of the hedge and also give a yield of fruit. Hybrid berries like tayberries and loganberries could also be used.

MEDICINAL HERBS

Many medicinal herbs grow on woodland rides, and it is possible to harvest your own medicines once you have gained some experience in basic herbalism and how to prepare the herbs. I use a few herbs for specific ailments. This knowledge is folklore and an important part of our cultural heritage, and using your own fresh and dried herbs is a healthy alternative to what is offered by the average pharmacist, and will save you money too. It is empowering to take responsibility for our own health and to use nature's abundant harvest. As with learning about mushrooms, a walk with a herbalist will reveal a new world growing on the extraction routes in the woodland. Remember some flowers/herbs are very rare and it is illegal to harvest them. Also as mentioned in relation to fungi, if you are uncertain about dose or identification, seek professional advice.

FUNGI

Woodlands and copses contain a large variety of different species of fungi, many of which are edible. My main edible species are forms of *Boletus* both the penny bun (*Boletus edulis*) and (*Boletus subtomentosus*) which grow throughout the coppice, along with chicken of the woods (*Laetiporus sulphureus*) and horn of plenty (*Cratellus cornucopioides*).

Wild fungi can generate a premium price if sold direct to restaurants. Learning to identify different fungi takes time to learn.

Opposite page

Top left:
Robert Hart's established forest garden, Wenlock Edge, Shropshire.

Top right:
Bilberry (our native blueberry) is a useful addition to an acid forest garden.

Below: Tayberries fruiting well in a brash hedge protecting coppice.

Centre left:
Self-heal is great for cuts and grazes is abundant along the woodland rides at Prickly Nut Wood.

This page

Above:
Chicken of the woods growing on a sweet chestnut stump, Prickly Nut Wood.

I have a simple rule: unless I am absolutely certain of the variety, I don't eat them! Walking through the woods with different people can shed light on a variety you have been uncertain of, or better still befriend an experienced mycologist.

Fungi can also be introduced to the woodland by inoculating freshly cut logs with fungal spores. Some of the most easily obtainable spores are from oyster mushrooms and Japanese shiitake mushrooms; both of which will bring in extra income for the woodland enterprise. Oyster mushrooms grow well on beech or birch logs and the shiitake do well on most hardwoods. The logs should be inoculated within a month of being cut in order to ensure other fungi have not established themselves first. Holes are drilled into the centre of the log and plugged either with a dowel inoculated with spawn or directly inoculated with spawn impregnated sawdust. The holes are then wax sealed to keep the spawn moist. The logs are stacked in a shady place, and once the mycelium has spread through the wood – usually taking between one to two and a half years depending on the type wood used – the logs can be shocked into early fruiting by immersing them in water for about 48 hours. Mushrooms should begin to appear about a week after the logs are removed from the water.

After the mushrooms have cropped, the logs need to be rested before reshocking, which can be done up to four times a year. The logs will keep producing until the wood decays. It is also possible to purchase young trees inoculated with mycorrhizal fungi which have a symbiotic association with the roots of the trees. It is possible to buy young hazel trees inoculated with black truffle (*Tuber uncinatum*). The truffles need a well drained chalky soil and will start producing from about five years after planting. With the demand for hazel coppice on the increase, this should be a perfect combination. (See Mushroom Spawn Suppliers in Appendix Eight.)

Opposite left:
A good harvest of Horn of Plenty fungi, Prickly Nut Wood

Left:
Shiitake mushrooms fruiting at Prickly Nut Wood.

Below (left and right):
Cornish top bar hive.

BEES

A few beehives in the woodland will help pollinate fruit trees and supply you with a good crop of honey. I have kept bees in the woods and provided they are situated where they will catch the early morning sun and are not overshadowed by trees, they will enjoy both the flowering plants on the woodland rides and the trees themselves. Pussy willow (*Salix caprea*) is one of the earliest bee fodders, and lime and chestnut flowers produce delicious honeys. I use the chestnut as the main honey crop and it produces a good flow of honey. Provided a design includes enough trees of good nectar flow, bee-keeping could form a useful economic part of a woodland enterprise.

Food From The Woods

Top bar hives are becoming more popular for the amateur beekeeper and can be obtained from: www.cornwallhoney.com

Bees are currently suffering a serious decline and any increase in bee fodder to help them re-establish numbers should be considered in woodland designs. Consideration should also currently be given to 'not taking' or minimal harvesting of honey. Honey is one of life's luxuries, but living bees are one of life's essentials.

MEAT

When I eat meat, I like it to be wild. So much of the meat that is readily available has been produced on chemically supported land systems and the animals themselves are pumped full of hormones and artificial growth stimulants which can be passed onto us as we consume the meat.

Many of the common species that a coppice worker has to keep controlled in order for the coppice to regenerate are fine sources of food. Rabbits are abundant and their meat is low in fat and when cooked as a stew over the fire it produces a rich flavoured source of protein.

Deer are on the increase in Britain and are the most destructive animals to woodland, especially coppice. With the introduced dwarf muntjac, they have a skilled team of all sizes, fleet of foot and able to jump great heights. They are beautiful creatures and deserve their place in the woodland, but as with all species (and maybe one day the same will be applied to ourselves), it is a question of balance. Large woodlands and estates employ a deerstalker, not to eradicate the deer population, but to control the numbers in a particular area of woodland to allow coppice regrowth or young trees to establish. The extra benefit is, of course, the wild venison.

Pheasants and wood pigeons are readily available in most woodlands and for those with diverse taste, grey squirrels which damage many trees through ring barking can make a tasty change to the menu.

Wild Boar

Wild boar are becoming more common in woodlands. Wild boar can be bred under licence (Dangerous Animals Act) within woodland and their meat fetches a high market price. The woodland needs to be securely fenced, but it is worth considering the damage they will do, especially to ground flora, before deciding on breeding them. Wild boar dig and root around and over time will cultivate the woodland floor as well as uprooting saplings.

Because of this, it is best to keep wild boar in woodlands with little ecological value, like poor quality plantations where the wild boars' ground preparation may help to encourage natural regeneration after they have left. Wild boar certainly live a better life than most reared animals, they have space to roam and their natural activities such as nest building can go unhindered.

Over the past decade, the population of 'escaped' wild boar free ranging in our woods has grown. In the Forest of Dean, the New Forest and parts of Kent, there are well established populations in the wild. Wild Boar, once again is becoming established in our countryside. The effect this may have on our woodland groundflora and bluebell populations only time will tell; but a management plan for numbers of wild boar will soon become part of many woodland plans and with that the benefit of another source of meat.

Opposite left:
Jointed rabbit ready for cooking.

Top right:
'Bushmeat Casserole', a Prickly Nut Wood favourite made from marinated squirrel.

Below right:
Wild boar sow with litter.

WOODLAND MANAGEMENT & THE LAW
(The Bureaucracy of Woodland Management)

Chapter Eight

There are a number of legal requirements concerning woodland which are important to be aware of. Most are purely commonsense and are safeguards to stop poor woodland management, although when I look at the sitka spruce plantations in Scotland I wonder why there is no legal control upon them. The majority are concerned with health and safety and the control of felling trees, and some are to safeguard the conservation status of a woodland. I have included planning law in relation to woodlands as I have experienced a number of grey areas surrounding forestry and planning. I have also included the English Woodland Grant Scheme at the end of this chapter, as these grants can be helpful in carrying out some of your woodland management objectives. Finally, you will find an example of a woodland contract (see Example Of Contract in Appendix Six). Aspects of planning law, felling licences, grants, etc. are all liable for review and possibly change, so contact the Department for Environment, Food and Rural Affairs (DEFRA) and the Forestry Commission for updates.

Felling Licences

You need a felling licence to fell growing trees unless the work you are carrying out falls into one of the categories listed below. Without the need for a felling licence you can, in any quarter, fell up to 5 cubic metres as long as no more than 70ft^3 (2m^3) are sold. A quarter is described as follows: 1 January to 31 March, 1 April to 30 June, 1 July to 30 September, 1 October to 31 December. So you could fell 177ft^3 (5m^3) on 31 December and 177ft^3 (5m^3) on 1 January and be within the law, provided you felled no other trees between 1 October and 31 March. You do not need a licence if any of the following conditions apply:

a) The felling is part of an improved plan under the Woodland Grant Scheme (see Grants at the end of this Chapter).

Above:
The mosses at Prickly Nut Wood are a key part to its SSSI designation.

b) The trees are growing in a garden, orchard, churchyard or public open space.

c) When measured 4ft3in (1.3m) from the ground, the trees are less than 3in (8cm) in diameter. If the trees are thinnings less than 4in (10cm) in diameter. If coppice or underwood less than 6in (15cm) in diameter.

d) The trees interfere with development permitted by planning legislation or work legally carried out by a public organisation.

e) The trees are dead, dangerous, causing a nuisance or are badly affected by Dutch elm disease. (I have still not yet met a tree which is causing a nuisance!)

f) The felling is done under an Act of Parliament.

If you need to apply for a licence you must obtain form PW11 from the Forestry Commission. In many cases it will be easier to apply for a Woodland Grant Scheme contract, as you may be able to obtain a management grant as well as permission to fell within the plans. Fines for felling without a licence can be in the region of £2,000 per tree, so you have been warned.

Tree Preservation Order

You cannot fell, top, lop or uproot a tree covered by a Tree Preservation Order. This is issued by the local planning authority and only by application through them can you affect change on a Tree Preservation Order. You may wish to put a Tree Preservation Order on certain trees. You can obtain a leaflet 'Protected Trees – A Guide To Tree Preservation Orders' from DEFRA.

Sites of Special Scientific Interest (SSSI)

If your woodland is a Site of Special Scientific Interest (SSSI) you will be given a list of any work that may damage the scientific interest of the land within an SSSI under Section 28 of the Wildlife and Countryside Act 1981. In England this will be administered by Natural England (NE), formally English Nature (EN); in Scotland by the Scottish Natural Heritage (SNH); and in Wales the Countryside Council for Wales (CCW). The present policy with EN is to agree a site management statement between EN and the owner which will allow management of the woodland to meet the needs of the scientific interest as well as the needs of the owner.

Felling licences or Woodland Grant Scheme contracts can apply to SSSI but they must also be approved by NE, SNH or CCW. SSSIs have derived a mechanism to safeguard rare habitats, but often at a cost! Many landowners have used the fact that if an area of land is of scientific interest, they can then prepare a management strategy that would damage the SSSI. This usually involves an annual payment by the Nature Conservation Body to make sure the new management does not go ahead, although occasionally a compulsory purchase is undertaken. The result often being that a landowner who does not care for the nature conservation of the land gets paid a regular and very often high value income to do nothing to his land!

Improved provisions for the protection and management of SSSIs were introduced by the countryside and Rights of Way Act 2000.

Other Designations

Other designations which may effect woodlands are:

AONB – Area Of Outstanding Natural Beauty.

AoSP – Areas of Special Protection, originally designated under the Protection of Birds Act 1954.

National Parks

Natura 2000 – European Union-wide network of nature conservation sites established under the EC Habitats and Birds Directives.

Ramsar sites – Designated under the convention on wetlands of international importance.

SAC – Special Areas of Conservation designated under the EC Habitats directive.

SPA – Special Protected Areas are classified by the UK government under the EC birds directive.

LNR – Local Nature Reserve.

NNR – National Nature Reserve. A search on any woodland should highlight any of these designations.

EC Habitats Directive Safeguarding European Protected Species
– The species that occur in English woodlands and for which it is an offence to disturb are:

All 17 species of bat
Dormouse
Great crested newt
Otter
Sand lizard
Smooth snake

The natterjack toad and some plant species, such as yellow marsh saxifrage, may occur rarely in woodlands and could be effected by forest operations.

The level of protection of these

species has been increased to comply with the EU Habitats Directive following a judgement in the European Court of Justice (ECJ).

The amended regulations include as an offence any damage or destruction of a breeding site or resting place. Previously if damage was 'an accidental result of a lawful operation' and reasonable precautions had been taken to avoid it, it would not have been an offence. Therefore there is a risk of woodland operators committing an offence if they have not carried out planned operations carefully, with the necessary checks and sought a licence where required.

The Forestry Commission website (www.forestry.gov.uk) gives presentations on implementing the habitats regulations and good practice in relation to the habitats regulations.

INSURANCE

The most important insurance for woodland work is Public Liability. Most land owners will ask for evidence of this before any work commences in woodland. The Forestry Contracting Association provide a comprehensive range of insurance for woodland work and at competitive prices through Algarth Insurance Brokers Limited. They have a special insurance for coppice workers which covers coppicing and woodland activities such as charcoal burning. It is well designed and meets the needs of most coppice workers. Employers' liability should be taken out if you employ people or train woodland workers, and you should consider an insurance on forestry machinery, especially if you purchase a mobile sawmill, for example. The Co-op offer some good value insurance packages.

HEALTH & SAFETY

Health and Safety at work is primarily common sense, however risk assessments will need to be carried out for different activities within the woodland. Risk gain should also be considered, there are some benefits to being exposed to risk for personal development. When felling trees you are likely at some point to need a chainsaw. To operate a chainsaw in woodland you

Above:
***Brown long-eared bat,
a common inhabitant of woodlands
in Britain and Europe.***

need to gain a Certificate of Competence. In England and Wales, this is through the National Proficiency Tests Council and in Scotland through the Scottish Skills Testing Service (for contact details for both, see Useful Organisations in Appendix Eight). You can take a training course usually of about four or five days and at the end you are tested on your ability to operate and maintain a chainsaw and fell trees within the health and safety guidelines. If you are considering using a chainsaw, these courses are highly recommended.

Alternatively, you can learn to use a saw under instruction on a one-to-one basis and then apply to an approved assessor to test your ability.

Woodlands & Taxation

If you fell a tree, sned it (cut off the branches) and cross cut the tree to size and sell it in the raw state as timber, the income earned is exempt from income tax. If you plank the tree or convert it into a finished product (even sharpen the end of a pole to make a stake) it becomes liable to taxation.

Planning Law & Woodlands

Brief History of Planning Law

In 1947 the Town and Country Planning Act came into being. Its mission was to ensure that the countryside was protected and saw all forms of development, with the exception of agriculture and forestry, as a threat to this aim. Planning law was consequently introduced to protect the countryside and in many respects it has succeeded, but it has also allowed farmers to build practically whatever they want. The notion that farmers have made the countryside what it is and have cared for and nurtured the land may have had some truth prior to 1947; but what was never envisaged was the destructive mechanisms of modern agriculture which were allowed to proceed undeterred. Farmers have been allowed to build vast silos, concrete and asbestos barns as opposed to oak timber frame, rip out ancient hedgerows, drain ecologically valuable water meadows, poison the land and waterways with agrochemicals and create deserts of wheat and food mountains, much of which is then dumped on the world market.

After joining the Common Market and applying the fixed prices of the Common Agricultural Policy, farmers have been subsidised handsomely to continue this damaging pattern. In the 1990 edition of the Town and Country Planning Act, it still follows the same guidelines of the 1947 Act. The only real change seems to place planning restrictions on smaller agriculture units of less than 12 acres (5 hectares). Reasons for this can be understood as there have been many cases of abuse of the planning system and pseudo agricultural enterprises being sold off for vast profits after planning permission has been gained. However, it also penalises the smallholder and previously landless person who works for years to save the money to buy an area of land to work only to be told, "If you could afford to buy more, we might allow you to live there."

Planning law has 'land use' categories for both agriculture and forestry. Although legislation has been similar for both, prior to the National Planning Policy Framework (NPPF), the last DEFRA Planning Policy Guidance Note 7, 'The Countryside Environmental Quality and Economic and Social Development' (Government guidance note upon which planners make decisions on rural development), made reference to the fact that it was unlikely to justify the need for new forestry dwellings as modern forestry rarely shows a need for people to reside upon the land. No such reference was made under agricultural dwellings. Although it would be possible to make a case that a forestry practice does not follow conventional modern methods of management and therefore a new forestry dwelling could be justified... etc., it brings into question what is agriculture and what is forestry.

Our earliest agriculture comes from forestry. The clearing of the wildwoods to cultivate land would have been the first step into agriculture as we now know it. This would have been linked to a hunter-gatherer lifestyle and harvesting of tree crops. As we have cleared larger areas and woodlands began to disappear, we have come to accept this land as agriculture. So agriculture grew out of forestry and therefore to try and separate the two (which are both activities of working with the land) seems unnecessary.

However, in planning law, the term agriculture is defined under a 1947 definition which only includes the use of land for woodlands where that use is ancillary to the farming of land for other agricultural purposes. This separates forestry as a distinct land use category in its own right. However, it lists under the definition for agriculture such practices as fruit growing and osier land, and osier land is certainly woodland. After all, it is grant aided by the Forestry Commission. It is consequently quite possible with a diverse permaculture designed woodland to argue that many aspects of the agricultural definition apply to a woodland. What this all shows is that the definition needs a clearer vision as land use is not always as black and white as the definition suggests it should be.

When To Apply For Planning Permission?

To many people who have acquired or who have access to land, the need and desire to reside on the land becomes quickly apparent.

Commonly Asked Questions:
Should I move onto the land before approaching the planning authority? Should I try to remain discreet, or should I declare my intention from the start?

The answer varies according to circumstance, but the following advice I hope gives a good balance in answering these questions.

1. Prepare a management plan for the land. This should include maps, access, short-term activities and long-term plans. Prepare a business plan for the land-based activities. This should include costings, cash flow projections and information on marketing strategies.

2. Consult with the local council about your plans, but first talk to the environment and conservation departments. This should help to gain support from some sections of the council before dealing with the planners. Interact and gain support at a local level.

3. Start working the land and move on in stages. Start simply and learn what your needs are and the needs of the land.

4. Deal with the planning department as and when they approach you, or it seems propitious to approach them. Here I am not suggesting that you move on to the land without planning permission, but you can legally use the 28 day rule or move on to the land as a seasonal forestry worker while establishing your business.

One important consideration worth remembering is that if you are established on the land and it is your only dwelling, you are in a stronger position to gain legal aid in the case of perhaps a High Court appeal if you end up in a complex legal exchange.

Two Key Planning Reference Documents

Before considering an application to the planning department, you are advised to have read and get a clear understanding of the latest *General Permitted Development Order* (currently 1995 version) and the recently introduced *National Planning Policy Framework* (2012) which now does away with the previous document, *Policy Guidance Statement No.7* and in particular Annex A of that document (which refers to agricultural and forestry dwellings). Both are available from either the DEFRA or Her Majesty's Stationery Office.

Every council provides a local development plan for their area and these can be read at the council offices. By reading this documentation you will gain an understanding of what the council is trying to achieve in the balance between development and conservation within the local area. It can be useful to quote relevant parts of this document in your application to the planning authority as it will outline how your application fits into the local development plan.

The Application

You may approach your planning authority prior to putting in an application to gauge their opinion on your proposals and to see what changes they might see as favourable in your application. There is a fee for this process.

Planning application forms are available from your local council offices. An application for planning permission will involve a fee, and the filling out of the relevant forms, plus maps and plans of the land and proposed developments. Your local planning authority is supposed to make a decision within eight weeks of receiving your application, but it often takes longer. During this time a notice will be placed at the edge of the land, and a mention in the local paper to make the public aware of the proposed application. A planning officer is likely to visit the land. This is a good time to obtain letters from organisations, individuals and charities who are supportive of your plans. They should be sent to the planning authority quoting the reference number of your application.

The planning authority then makes a recommendation to permit or refuse the application. Its decision, along with its reasons, is put before the Development Control Committee made up of elected councillors who will discuss and vote on the decision. If there is a lot of debate, they may defer their decision until they have carried out a site visit. The Development Control Committee can overturn the planning authority's recommendation to permit or refuse permission, but in practice they rarely do.

The Appeal Process

If your application is turned down you can appeal to the Secretary of State for the Department for Environment, Food and Rural Affairs (DEFRA). Your appeal must reach the Planning Inspectorate within six weeks of the date the local planning authority refused your application. This means filling in appeal forms available from: Tollgate House, Houlton Street, Bristol BS2 9PJ or in Wales from:

The Planning Inspectorate, Cathays Park, Cardiff CF10 3NQ.

An appeal can be resolved by any of the following three options:

1. Written Representations
You present a written version of your case and the planning authority presents a written version of theirs. An inspector appointed by the DEFRA makes a decision. This process gets the quickest result, but it is the least recommended or fair of the appeal choices, as it relies on good literacy.

2. Informal Hearing
This involves presenting the written case as in option one, but allows a hearing, where the DEFRA appointed inspector will encourage questioning and discussion with yourself and the planning authority and you will be allowed to explain your case in further detail at the site visit which is usually at the end of the hearing.

3. Public Enquiry
This involves presenting the written case as in the other options, but involves a public enquiry which acts a bit like a Magistrates Court with the DEFRA appointed inspector presiding. With this option you present your case, the planning authority presents theirs and you have the option to cross examine the planning authority and they the option to cross examine you. Members of the public can attend and witnesses can be called to validate your case. A site visit is conducted but you cannot add further information to your case during this visit.

The advantages of the public enquiry are that it gives you or your legal

Above:
First forestry building at Prickly Nut Wood.

representative an opportunity to present your case prior to the site visit, and it gives you time. Public enquiries can take up to a year to organise which gives you another year on the land to show the genuine intention of land use.

An inspector is chosen from a 'respected' position in society. This is often a solicitor, barrister or engineer; one strange quirk is that to be an appointed inspector you have to hold a full driving licence! One of the difficulties with the appointed inspector system is that however impartial they try to be, they are just one individual. This person has their own patterns of thinking like you or me and however open minded and well trained they are, their decision may be affected by their viewpoint on the countryside. A panel would certainly give a fairer decision and avoid situations like the one I encountered. At an enforcement notice appeal, the Inspector compared the need to reside on the land to oversee charcoal burning activities as being the same need as to reside on the land to convert pigs to sausages! There was obviously a lack of understanding by the inspector about the history of the countryside and the essential need to supervise fire in a woodland. The decision of the inspector can very occasionally be overruled by the Secretary of State in exceptional or highly controversial cases.

Before deciding whether to appeal you should accept that in most cases you will need expert advice from a forestry consultant to evaluate your management plan and someone with a legal background in relation to planning law. This advice can be expensive. The actual costs of the appeal itself are usually met equally by both parties; in other words they pay for their own costs unless one party can prove that the other has acted 'unreasonably' from a legal definition of 'unreasonable' under planning law.

If the appeal fails, you then have the option of taking the process on to the High Court. This is likely to be very expensive unless you can obtain Legal Aid. You will need expert legal representation at this level. If you lose at the High Court you can appeal to the European Court of Human Rights provided you have referred to Article 8 of European Human Rights throughout your appeal process.

Application For A Non-Residential Building

You can build a forestry building for forestry use (not a dwelling) on a forestry enterprise of any size. This comes under part 7 of the 1995 General Development Order. You need to notify your planning authority of your intention and the size, position and type of building to be constructed. This is called a 'prior notification procedure'. The planning authority has 28 days to reply to your notification. If they do not you automatically have planning permission. The planning authority may dispute the size of your building, the look of it or position of it, but you have permitted development rights to build a building, you may have to reach a compromise on the size, look and position. The planning authority may refuse your permitted development rights if you propose a large building on a three acre woodland site. They could argue that such a building would be out of proportion with the forestry enterprise.

Application For A Forestry Dwelling

Whatever size the woodland is you will, in most cases, need to apply for planning permission. (It is still possible to get a Certificate of Lawful Use after four or ten years depending upon circumstance.) The larger the woodland operation is the more likely planning permission will be granted (see above).

With the Department for Communities and Local Government's new National Planning Policy Framework (NPPF) taking over from Planning Policy Statements (PPSs) and therefore the removal of PPS7 and in particular Annex A, planners currently find themselves in an uncertain position as to how to decide the merits of a forestry or agricultural dwelling application. In the past there were clear guidelines. The two essential tests, 1) economic viability of the enterprise and 2) essential need to reside on the land, and guidance on how to measure these tests. These have now been removed. The NPPF still advises local authorities to avoid new isolated homes in the countryside unless there are special circumstances such as the essential need for a rural worker to live permanently at or near their place of work in the countryside. There are two difficulties here that the NPPF leaves local authorities; the first is that 'rural worker' is not defined and the second is no details are given on how to assess 'essential need'. It seems the Government is leaving local authorities to determine what this means locally and incorporate this into their new NPPF compliant local plans.

It seems that until these new local plans are in place, local authorities are falling back and adopting the now defunct Annex A from PPS7 as their guidance for assessing applications. One example is Colchester Borough Council who have adopted into their local plan framework a 'Planning Guidance Note for Rural Workers'. This is set out to be an advisory note to applicants wishing to apply for a rural workers dwelling in the countryside. The document reverts back to using the criteria of Annex A

Above:
Caravans can be of many types.

from PPS7. A number of other local authorities have taken a similar position adopting Annex A until the government provides further guidance. Meanwhile, Government under Lord Taylor is reviewing all the practice guidance notes in order to create a new web-based government guidance resource by July 2013. Annex E to PPG7 (1997), which deals with permitted development rights and the determination procedure, is the only remaining guidance on forestry matters, and Lord Taylor proposes that this should be cancelled now and redrafted later. There is no proposed redraft of Annex A of PPS7.

So where does this leave a forestry worker applying for a permanent dwelling?

It now seems it may no longer be necessary to apply for a temporary three or five year permission in order to prove the economic viability of the forestry enterprise. (Although it is possible that local planning authorities could keep temporary permissions without conflicting with the NPPF.) With a good business plan and residing on the forest land within a caravan as a seasonal forest worker, you should be in a far stronger position to apply for a permanent dwelling rather than a temporary dwelling. The criteria of essential need is still in place and that will be the challenge for any planning application. I anticipate that most planning authorities will fall back on annex A of PPS7 as they await government guidance that may never come, and it will be decisions at planning appeals and in the courts which will challenge the reliance on this now defunct annex. There will no doubt be some interesting cases that challenge the wording of 'rural worker' and open up its definition.

Your application will need to be accompanied by maps, design of the dwelling, a business plan to show the economic viability of the enterprise, incorporating your essential need to reside there, a forestry management plan and of course the fee. The application will then follow the same proceedings as a forestry building through the local planning authority.

The decisions on all planning applications are made taking into

Woodland Management & The Law 151

account the local planning authority's local plan. Having the knowledge of this and quoting relevant extracts can help frame an application, particularly in this time of change with the new NPPF. In addition to these, if the application is in a sensitive area, other bodies can voice an opinion. In my application for Prickly Nut Wood, for example, English Nature were consulted as the wood is a Site of Special Scientific Interest, as were the Sussex Downs Conservation Board because it is an Area of Outstanding Natural Beauty, and also consulted, of course, was the Parish Council.

Temporary Permitted Stays On Forestry Land
The 28 Day Rule

Under the General Development Order Part 4(b) 1995, you are entitled to stay in a woodland in a caravan for a maximum of 28 days in a calendar year without planning permission.

Seasonal Forestry Workers

Under the General Development Order Part 5, permission is granted for the use of land as a caravan site in circumstances specified in paragraphs 2-10 of Schedule 1 of the Caravan Sites and Control of Development Act 1960. Paragraph 8 of this act refers to forestry workers and states that a site licence (in this case referring to the need to apply for planning permission) "shall not be required for use of land and the accommodation during a particular season of a person or persons employed on land with the same occupation, being land used for the purposes of forestry (including afforestation)."

In lay person's language, you do not need planning permission to stay on the land in a caravan for a season to work the woodland or plant trees within the forest or to create a new woodland on barren land. It seems there is as yet no definition of the word 'seasonal' as entered in planning law for this situation, so the justification of what amounts to a forestry season is very much up to your interpretation and that of the local planning authority. There is a definition of the word 'caravan' as defined in the 1960 Act which is "any structure designed or adapted for human habitation which is capable of being moved from one place to another (by being towed or transported on a motor vehicle or trailer) and any motor vehicle so designed or adapted, that does not include: a) any railway rolling stock which is for the time being on rails forming part of a railway system or b) any tent." Sadly, low impact structures like yurts or benders which are far more suitable for a woodland than a towed structure, come under the classification of tents. I have tried to argue the case that a yurt can be a caravan, as they are traditionally moved from one place to another, but the planning inspector decided that they came under the definition of tents.

Please refer to Appendix Four, Planning for Sustainable Woodlands, for suggested changes to present planning policy and woodlands.

Planning Experience At Prickly Nut Wood

In January 1995, I received a letter from Chichester District Council referring to a complaint that I was dwelling at Prickly Nut Wood in a structure of hazel poles

Opposite left:
The new roundwood caravan design for seasonal forestry workers at Prickly Nut Wood.

Below:
The Woodland House at Prickly Nut Wood.

Woodland Management & The Law 153

covered in tarpaulin (a bender). I met with the development control officer at Prickly Nut Wood, explained that I was staying on the land as a seasonal forestry worker and showed him the forestry activities I was undertaking, including charcoal burning and expected to hear no more of the matter. On 22 March 1995, I received a letter from Chichester District Council suggesting I removed the structure and vacated the land within two months. There then followed a flow of letters between myself and the planning authority, resulting in a build up of mutual mistrust between both parties. At the planning authority's Development Control Committee meeting of 22 August 1995, the committee decided planning permission was needed and enforcement notices should be served requiring the removal of the tarpaulin tent/structure from the land with a compliance period of six months. At this stage the bender had evolved into a 14ft (4.3m) diameter sweet chestnut yurt. I felt I had been somewhat railroaded by the planning authority at this meeting as there were many inaccuracies in the documentation given out to the Development Control Committee including a key page of a letter which was omitted in the documentation, a page that stated my livelihood was dependent upon using a shelter when working seasonally on the land. I complained to the Local Government Ombudsman that I had suffered injustice caused by the local planning authority's maladministration.

I then received a letter from the Chief Planning Officer explaining that the page of the letter was omitted in the copying process due to incorrect copying and that my situation would be raised again before the Development Control Committee on 3 January 1996. There was subsequently considerable discussion at that meeting and the decision for issuing an enforcement notice was upheld. Enforcement notices were issued and I appealed against the decision but could not convince the Appeal Inspector that a yurt was a caravan, despite historical evidence to the contrary (both of Mongolian yurts being moved on carts and of charcoal burners and bodgers living in huts of turf and wood in England). I had to remove the yurt and instead I installed an aluminium mobile home purchased second hand from a holiday camp. This was moved into the woodland with great difficulty and some damage to access tracks and trees. I applied for a temporary planning permission. This was granted for three years, with a personal condition limiting occupation to myself and any resident dependants. In conclusion, the strict application of the definition of a caravan in Section 29 of the 1960 Caravan Sites and Control of Development Act meant that a biodegradable self-build yurt was unacceptable as accommodation for a seasonal forestry worker, while an aluminium mobile home, which is totally out of place and impractical in the woodland location, was deemed acceptable provided a temporary planning permission was applied for.

Three years later, in 2001, I applied for a woodland cabin as a permanent dwelling. This was granted with personal conditions tying the use of the cabin to myself and to the management of the coppice and charcoal burning activities. In 2012, the conditions were relaxed to allow the cabin to stay and the personal condition was removed to allow someone working in forestry at Prickly Nut Wood to occupy the cabin.

This means in the long term when I am no longer able to work the woods, another person can continue to live in the cabin provided they continue to work the woods.

A Word of Caution

Think carefully before engaging in what could end up being a long, drawn out planning case. Hopefully in the future there will be an easier route for those of us who are trying to live sustainably on the land. I have seen many people suffer under the stresses and strains of a drawn out planning scenario – myself included. Do remember that you may be living for a long period of time, wondering whether you will be evicted from your home. Some people are more adapted to dealing with the stress and uncertainty than others; some people live with it all the time and have an accepting attitude towards the 'pressure'. I found the process very exhausting and I was fortunate to have some caring and listening friends as support. It also used up over 40 working days of my life (40 days I should have been using to look after the woodland and earning a living), and that cost me around 10 percent of my annual income.

Try to remember that the local planning authority is trying to understand what are unusual applications and sort out the difference between genuine people who need to live and work in the woodland and those who may have intentions towards land speculation.

I hope that this section makes the planning process a little easier to comprehend; I know how much I could have done with a simple guide before embarking on three years of planning stress. So weigh up the odds and be prepared before entering the arena.

To get a fuller picture on planning issues, I highly recommend reading Simon Fairlie's *Low Impact Development – Planning and People in a Sustainable Countryside* (see Bibliography in Appendix Seven) and keep up to date with your planning authority's local plan.

GRANTS

Grants change and are updated at regular intervals, so please check with the contact numbers for any possible changes.

The English Woodland Grant Scheme is the Forestry Commission's scheme to encourage people to manage and plant new woodlands all over Great Britain (see Useful Organisations in Appendix Eight). There are two main schemes, the first is the Woodland Grant Scheme and the second the Farm Woodland Premium Scheme which includes the Woodland Grant Scheme but gives extra subsidy for the conversion of agricultural land to woodland.

The English Woodland Grant Scheme

The English Woodland Grant Scheme (EWGS) is a contract drawn up between the Forestry Commission and the woodland owner/manager. The contract is for five years and includes a list of operations of work that will be carried out in the woodland during the five year period. There will be an outline plan for the next 20 years. These may include felling, tree planting and conservation management. There is no need to apply for a separate felling licence if you are under a Woodland Grant Scheme contract.

Grants for New Planting of New Woodlands Under the Woodland Creation Grant

Conifers: £1,200 per hectare
Broadleafs: £1,800 per hectare
Special broadleafs £700 per hectare

The Forestry Commission uses a scoring scheme to rank applications and distribute funds. For more details of the EWGS and scoring scheme visit the Forestry Commission web site www.forestry.gov.uk/ewgs

or write to Forestry Commission England, 620 Bristol Business Park, Coldharbour Lane, Bristol BS16 1EJ and ask for an EWGS application pack.

Other grants available for woodlands through the EWGS include:

Woodland Planning Grant (WPG)

To pay for preparation of plans that both assist with management of the woodland and meet the UK Woodland Assurance Standard (UKWAS).

Woodland Assessment Grant (WAG)

Gathering of information and surveys to improve management decisions.

Woodland Regeneration Grant (WRG)

To support natural regeneration and replanting after felling where this produces desirable change.

Woodland Improvement Grant (WIG)

To support work in woodlands to create, enhance and sustain public benefits. This can include grants towards improving biodiversity, removal of rhododendron, deer fencing, recoppicing derelict coppice, improving public access and for forest schools.

Woodland Management Grant (WMG)

Contribution to additional costs of providing and sustaining higher quality public benefit from existing woods.

OTHER GRANTS

Environmental Stewardship Grants

This builds on the former countryside stewardship grants and has a range of levels from entry through to higher level stewardship. Can be useful funding where woodland is owned/managed in conjunction with other land.
www.naturalengland.gov

Crofting Grants (Scotland)

The Crofting Counties Agricultural Grant scheme helps support the viability of a crofting enterprise.
www.scotland.gov.uk
www.crofterscommission.org.uk

Tree Council

The Tree Council gives grants for planting of trees in schools and by community groups.
www.treecouncil.org.uk

Royal Forestry Society

The best information on a wide range of grants available for forestry can be found via the following link to the Royal Forestry Society.
www.rfs.org.uk/about/grant

CONTRACTS

Most agreements have been made verbally or by a shake of a hand between woodland workers and land owners, but in recent times, cases of disagreement and difficulties have arisen from both parties. Hence the need for a contract to lay down who is responsible for what and to provide the opportunity for mediation in the case of unresolved difficulty.

The example contract in Appendix Six is the one I drew up for the Sussex and Surrey Coppice Group and could easily be adapted for non-coppice woodland agreements. In this case, it is for the use of members of the Surrey and Sussex Coppice Group and therefore the group acts as a mediator in the event of difficulties. The Forestry Contracting Association also has a contract for members and will also mediate in the case of difficulties.

THE FUTURE

Chapter Nine

As we look to the future of woodland management many of the keys are clearly visible in the past. Traditional management practices linked to local marketing and a raising of awareness in consumer consciousness will all be key factors in ensuring the sustainable management of our woodland resource. A closer connection of people to their local woodlands with the help of improved public access is just one part of raising awareness. Certification of woodlands and produce will be another, as consumers start to look closer at what they are buying.

The importance of buying local products could be highlighted by a 'timber miles' campaign to look at the distances a bamboo cane, a bread board or a plank of wood have travelled. Timber is a heavy product to transport and its carriers are large fossil fuel consuming and polluting vehicles which also add unnecessarily to road congestion. A travel tax could be introduced on imported timber and imported timber products, the cost being in relation to the distance travelled. This would make an 11 pound (5kg) bag of Indonesian rainforest charcoal cost, for example, £10 and it would no longer be a cheap alternative to the sustainably produced local product. This would be a case of using taxation to reduce travel, reduce the use of fossil fuels and prevent rainforest destruction. Income from the taxation could be channelled to help set up local 'wood stations' with finishing equipment like planers, routers and moulders, etc. Local people could collect floor boards direct from their local wood station with the knowledge of exactly which woodland they had come from. These wood stations would be of benefit to small-scale woodspeople who could take their timber and convert it into the desired finished product with minimal transport costs. The finished products and buildings would be a testament to future generations of local distinctiveness and sustainability.

Education at all levels will shape our attitude towards woodlands and in particular the need for training a new generation of foresters in sustainable forestry. The emphasis must switch from the industrial clearfell model to selective forestry and from short-term monetary economics to sustainable economics evaluating all the energy costs in a forestry operation and above all the health of the woodland.

Above:
Charlton Orchard's stall at the Glastonbury Farmer's Market, Somerset.

The differences between present forestry practice and sustainable woodland management

Present forestry practice	**Sustainable woodland management**
Trees are viewed as timber	Trees are an integrated part of a diverse ecosystem
Industrial management – use of pesticides, fertilisers and large machinery	Sustainable management: no chemicals, small-scale machinery and biological resources (horses to extract)
All trees are of similar age	Trees of all ages (seedlings to veterans)
Short length rotations	Long rotations (leaving veterans)
Clearfell is main management practice	Selective felling and coppice management
Monoculture plantations are favoured	Diversity is favoured
All forest is managed	Areas are left for their own evolution
Timber is the only forest product	Woodland offers diversity of products, timber and non-timber, recreation etc.
Traditional forest knowledge is seen as outdated and ignored	Encouragement of traditional knowledge, coppice woodmanship, craftwork, herbalists
No sacred places or woodland celebration	Woodland offers environment for celebration, contemplation and spiritual growth
Economics based on 'short-term monetary thinking model'	Economics based on long-term sustainability of the whole woodland ecosystem, its diversity and multifunctional yield
Forest workforce trained as timber harvesters	Woodland training through observation, traditional knowledge, and understanding of the whole ecosystem
Timber transported long distances	Timber and non-timber produce sold locally
Management plans drawn up after brief visit	Management plans drawn up after long-term observation through different seasons
Forest designed around machinery operations	Woodland design based upon permaculture principles
Management has high capital investment (large machinery)	Management is people (labour) based and locally sourced

Above:
100 Devon varieties at an Apple Day celebration.

Opposite right:
The author, Paul and Neil getting set up for the annual Lodsworth apple pressing day.

Localisation

The link between sustainability and the importance of localisation is one of the greatest difficulties we all face as we try to come to terms with living in a more sustainable way. We have become so used to having exactly what we want, when we want it, to a degree where many children growing up have no concept that a particular apple has a particular season when it is available. More frightening still is that the natural process of growing food is being bred out of us – we are producing at least the first generation of children that do not know how to cultivate the land and grow food! This may all seem unnecessary as our supermarket shelves are so well stocked, but it takes only a small jolt in our fragile food distribution system to have a large knock-on effect. I was in the South of France in December 1996 when the French lorry strike was at its peak. Roads became blocked and petrol became increasingly unavailable and the fruit and vegetable market at Toulouse (the equivalent of Covent Garden in the UK) shut down as there was no produce. The dispute was resolved before the situation became worse, but it was a clear example of how unsustainable we have become when we rely on obtaining our needs from a great distance.

The need to obtain our basic requirements locally, such as food or wood, and reduce the distance they travel are key issues we need to address now. Local authorities have the difficult task of introducing Local Agenda 21 strategies into an already bureaucratic system. Low impact sustainable development will consequently become a growing issue in the future of land use whether it is agricultural or forestry (if it continues to be necessary to see them as different disciplines). Imagine, for example, a woodland enterprise which manages a woodland in an environmentally sensitive manner; markets all produce locally; creates opportunities for rural employment; minimises the use of fossil fuels; recycles all waste and uses rainwater harvesting (as opposed to draining local reservoirs) and requires only a low impact dwelling, made from local resources... Are the local planning authorities going to be able to continue to argue that this is unsustainable when their own offices may contain timber that has travelled half way round the world? These are the difficulties we face as we try to meet the needs of the present without compromising the opportunities for future generations to meet their needs.

Eleven Woodland Recommendations

1. Long-term planning/thinking must accompany any new planting plans and species choice should be as diverse as possible.

2. Localised marketing should be promoted and encouraged.

3. Small low impact forestry development should be encouraged by local authorities provided they are safeguarded against land speculation.

4. The sale and marketing of non-timber woodland products should be encouraged, e.g. nuts, fruits, woodland honey.

5. An environmental travel tax should be introduced on imported timber. The cost of the tax should be in relation to the distance the timber has travelled.

6. Small-scale community wood stations with wood finishing equipment should become an integral part of any local woodland economy, supporting the small-scale woodsman who wants to convert (for example) overstood coppiced ash or alder to floorboards.

7. Improved public access to woodland, wood pasture and commons to become an urgent priority.

8. More wildlife corridors and hedges should be created to link existing woodlands.

Martin Lester

9. Allow wilderness in carefully selected areas, where woodland could be left to evolve without human intervention.

10. Parish woodland maps and inventories should be drawn up, with records of species, type of management and tree preservation orders should be established on any chosen trees.

11. Increased biosecurity measures should be put in place. This should cover entry to the country of plants, seeds, timber and humans. (all of which can be carriers of plant pathogens).

Celebration

The move towards sustainability in woodlands and in every aspect of our lives can only come from within each and every one of us. The motivation towards sustainability is the easy part, it is having the knowledge and the practical skill to implement the theoretical that will ensure our grandchildren enjoy the beauty of British woodlands. We can no longer expect to be subsidised to manage the land sustainably. True sustainability will come from us earning a living from working the woods, not designing management plans around Forestry Commission grant schemes, we must make the changes in our lives and in our attitude to what we consume and how we treat the planet. In this difficult time of change, trees can play a vital role as they have so often throughout history. The revival of tree celebrations can help make us aware of the local variety and resource of our local woodlands. Across the country particular villages and towns have days of tree celebration, like Great Wishford in Wiltshire on 29th May, where villagers enter the nearby Grovely Wood to cut oak boughs which are brought back into the village and placed in front of the houses. One large bough is decorated and hoisted to the top of the church tower to the chant of 'Grovely, Grovely'. These festivities ensure that the parishioners of Great Wishford keep their ancient commoners rights to gather firewood from Grovely Wood. These unique days, often with roots in ancient festivals, celebrate the beauty and abundance of our trees and plants. Tree dressing ceremonies (usually now on 29th May) and Apple Day Celebrations (usually 21st October) are becoming more common across the countryside.

These events give children and adults a closer connection to particular trees and their produce, and generally make us more aware of what is available within our locality. They also make us slow down enough to spend a day celebrating the abundance of all that is around us in ...

The Woodland Way.

The Future | 161

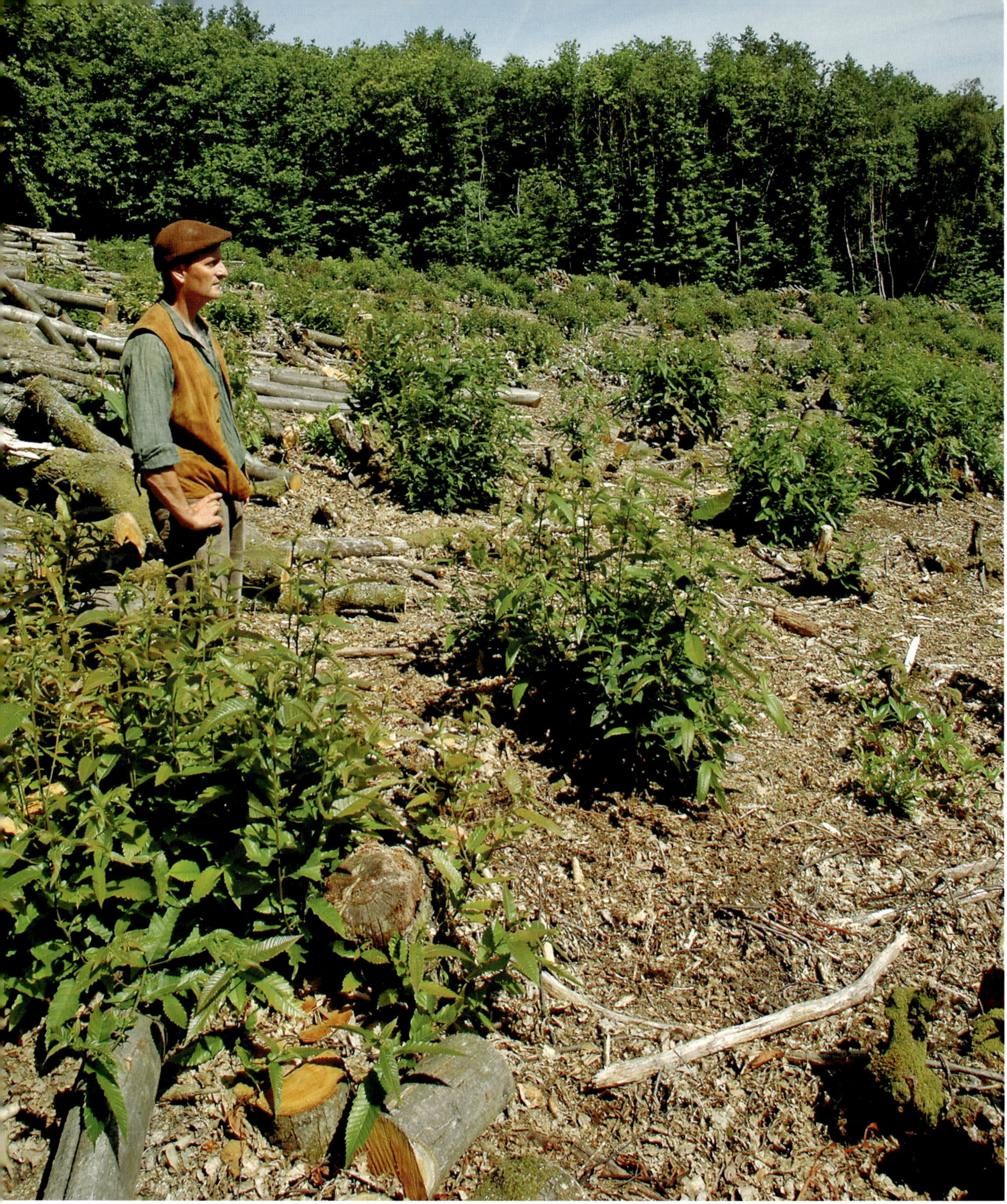

John Adams

EPILOGUE

The footpath cuts right through the garden and before it enters Paul's field, I collect walnuts from the gnarled tree that overhangs the path. The field was resown three years ago and some wild species are starting to reinhabit the sward. The path climbs past two old oaks which have seen more changes than I will ever know. The gate is a modern one and has a pronounced metallic clang as it shuts, enough to alert the farm to my presence. The young calves are resting in the shed, they all have names, steam rises from their velvet nostrils, another gate and past the barn.

The unfriendly hiss of a pylon makes me quicken my pace – that powerline runs from Dungeness to Exeter. It snakes through the valleys, yet when it was planned, it was going to be placed on top of the South Downs – fortunately the campaigners won that day. I turn and look back, the Downs look glorious today, their shape curvaceous and soft, but for me it's onwards – the chestnut beckons ahead.

Through the clover ley, whose sweetness lies in jars on my larder shelf, I'm up on the escarpment now and the footpath clings to the woodland edge; a little further then I descend. There is no pathway now – a veteran chestnut spirals upwards – and then deep into the copse. Chestnut stools look similar to an untrained eye, but there are subtle differences which make them familiar and help me to find my way home. Past the birch I tapped in the spring whose wine is almost ready and down the bank and over the footbridge – I will rebuild it in the spring – but my feet instinctively choose the solid pole and I emerge into the clearing.

Smoke rises gently, reassuring me the fire has stayed in; my kitchen is still outside, it will be another month before I take it in. The old iron kettle is ready to pour. I sit on a chestnut bench. I know the tree well that I cut it from and look up into the chestnut trees, not long now until they ripen, the ground still has remnants of the previous year's crop.

My mind wanders back over the passing seasons and I remember the midsummer picnic and the children running barefoot to the swing. They soon learn about the prickles, it's how the wood got its name. I remove my boots and caress the earth with my feet, the soil is cooler now, the woodcock flies over with his familiar 'rurrp'. I build the fire up and glance across to the ancient oak, soothing words enter my head, 'I know this place, I belong'.

Ben Law
Prickly Nut Wood
2001

Appendix One — USEFUL TREES, PLANTING CONDITIONS & PRODUCE

Key

Ac	acid soils	H	prefers heavy soils	N	neutral soils	St	stooling
Alk	alkaline soils	Hf	high forest	P	pioneer	Su	suckers
C	coppice	L	layering	Po	pollard	T	tolerant of most soils
Cu	cuttings	Li	prefers light soils	Rcu	root cuttings	W	likes wet soils
D	likes dry soils	Ltl	lower tree layer	S	seeds		
E	edge species	M	likes moist soils	Sh	shade tolerant		

Common Name	Latin Name	Soil	Propagation	Management	Timber Produce	Non Timber Produce
Alder buckthorn	*Frangula alnus*	N-Ac	S	C	excellent charcoal	brimstone butterfly food
Alder, common	*Alnus glutinosa*	W N-Alk	S	C	charcoal, underwater foundations, river protection, clog soles, brush heads, tool handles, workshop floorboards	dye from bark, nitrogen fixing, sacrifice lure for woodworm
Alder, grey	*Alnus incana*	N T H	RCu	Su	similar to common alder	nitrogen fixing, windbreak
Ash	*Fraxinus excelsior*	N-Alk	S	C Hf	gate hurdles, tool handles, hay rakes, steam bending, barrel hoops, tent pegs, furniture, floorboards, firewood, charcoal, walking sticks, necklaces, oars, hockey sticks, cues, yurts, long bows, ladders etc	chutney from seeds, canker growths used for firelighters

Common Name	Latin Name	Soil	Propagation	Management	Timber Produce	Non Timber Produce
Aspen	*Populus tremula*	M H N-Ac	Cu Su St	Su	firewood, pulp	edible inner bark
Beech	*Fagus sylvatica*	N-Alk	S	Po Hf Sh	furniture, firewood, spoons, chopping blocks, flooring, steam bending	nuts produce oil, edible leaves
Birch, silver	*Betula pendula*	T Li	S	C P	besom brooms, horse jumps, bobbins, firewood, charcoal, internal furniture, bark veneer	soil enrichment through leaves and twigs, sap wine, flammable bark, dye plant
Birch, downy	*Betula pubescens*	T M	S	C P	same as silver birch	same as silver birch
Blackthorn	*Prunus spinosa*	T	L Su St	Su	walking sticks	sloes for wine/gin, butterfly food
Box	*Buxus sempervirens*	Alk	Cu S	Ltl C	chisel handles, carving, door knobs, printing blocks, pestles	
Bullace	*Prunus domestica* ssp. *insititia*	T	S	E	turnery, furniture, firewood	edible fruit, dye
Cherry, bird	*Prunus padus*	N-Alk	L S Su St	Su	turnery	
Cherry, wild	*Prunus avium*	T	L S Su St	Su Hf	wood turning, furniture, veneer	edible fruit, resin used as chewing gum
Chestnut, sweet	*Castanea sativa*	Ac L	S	C Hf	hop poles, fencing posts, pallings, furniture, buildings, shingles, steam bending, walking sticks, yurts, gate hurdles, ladder rungs, floorboards, charcoal, firewood (woodburner), faggots, trugs	nuts, bee tree

Useful Trees, Planting Conditions & Produce

Common Name	Latin Name	Soil	Propagation	Management	Timber Produce	Non Timber Produce
Crab Apple	*Malus sylvestris*	T	S Su St	C	carvery, turnery, mallets, firewood, smokery	fruit used for jams, wines, verjuice, can be used as a rootstock for grafting
Dogwood	*Cornus sanguinea*	Alk W	Cu L Su St S	C Su	walking sticks, skewers, artists' charcoal	edible oil from seed
Elder	*Sambucus nigra*	N-Alk	S Cu	C	pipes, whistles, carving	wine, tea, medicinal uses
Elm, English	*Ulmus procera*	T	St Su	Su	furniture, turnery, mallet heads, building, floor boards, coffins, seats	
Elm, wych	*Ulmus glabra*	T H	S	C		bark used for seat weaving
False Acacia	*Robinia pseudoacacia*	T Li	S Su	Su	fence posts, tool handles, turning, walking sticks	flowers used for perfume, nitrogen fixer, bee tree
Guelder rose	*Viburnum opulus*	N	Cu S	E		bark used to treat cramp, berries cooked for jam
Hawthorn	*Crataegus monogyna*	T	S	C	firewood, charcoal, rustic furniture, walking sticks	edible leaves, berries for wine, jam
Hazel	*Corylus avellana*	N-Alk H	S L St	C	hurdles, thatching spars, hedging stakes, walking sticks, rustic furniture, benders, basketry, faggots, charcoal, firewood, pea and bean sticks etc.	edible nuts and leaves, soil enriching
Holly	*Ilex aquifolium*	N-Alk D	S L St	Ltl Su C	turnery, carving, furniture, riding crops, walking sticks, firewood	

Common Name	Latin Name	Soil	Propagation	Management	Timber Produce	Non Timber Produce
Hornbeam	*Carpinus betulus*	T H	S	C Sh	wooden cogs, firewood, charcoal	
Juniper	*Juniperus communis*	N-Alk	S Cu	Ltl	firewood, temporary stock fence	flavouring gin, medicinal berries, incense
Larch, European	*Larix decidua*	N-Ac	S	Hf	durable softwood, fencing, building	edible inner bark, dye from needles
Lime, common	*Tilia x europaea*	T	Rc St	C	turnery, carving, bee hives, furniture, bark used for seat weaving	edible leaves, tea from flowers, bee tree, rope from bark
Lime, large-leaved	*Tilia platyphyllos*	T	Rc St	C	same as common lime	same as common lime
Lime, small-leaved	*Tilia cordata*	T	Rc St	C	same as common lime	same as common lime
Maple, field	*Acer campestre*	T H	S	C	turnery, musical instruments, carving, firewood, furniture, spoons, bowls	wine from sap
Medlar	*Mespilus germanica*	T M	S	E	firewood, carving	edible fruit when bletted
Monkey Puzzle	*Araucaria araucana*	T	S	Hf C	construction timber	high yield of nuts once mature, not self fertile
Mulberry	*Morus nigra*	T	S	E	furniture, carving	edible fruit
Oak, English	*Quercus robur*	T	S	Hf C	buildings, furniture, floor boards, barrels, swill baskets, shingles, fencing, firewood, trugs, charcoal	flour from acorn, bark for tannin
Oak, sessile	*Quercus petraea*	T M	S	Hf C	same as English oak	same as English oak

Common Name	Latin Name	Soil	Propagation	Management	Timber Produce	Non Timber Produce
Scots Pine	Pinus sylvestris	N-Ac L	S	Hf	construction timber	resin
Rowan	Sorbus aucuparia	N-Ac	S	C	walking sticks, carving, rustic furniture, firewood	fruit for jam, wine
Plum, wild	Prunus domestica	T	S	E	turnery, furniture, firewood	edible fruit, dye
Spindle	Euonymus europaeus	N-Alk	S	C	spindles, skewers, knitting needles, artists' charcoal	medicinal fruit
Sycamore	Acer pseudoplatanus	T	S	C Hf	turnery, clogs, kitchen utensils, veneer	wine from sap
Whitebeam	Sorbus aria	N-Alk	S	C	walking sticks, firewood	
Wild service	Sorbus torminalis	N-Ac H	S Su St	Su Hf	musical instruments, furniture, charcoal, firewood	edible fruit, wine
Willow, almond	Salix triandra	T W	Cu	C	river protection, wattles, thatching spars, baskets, cribs, artists' charcoal, living structures	land reclamation, giberalic acid for plant rooting, rope, teething rings, bee tree
Willow, bay	Salix pentandra	T W	Cu	C	same as almond willow	same as almond willow
Willow, crack	Salix fragilis	T W	Cu	C Po	same as almond willow, but not baskets	same as almond willow
Willow, goat	Salix caprea	T W	Cu	C Po	same as almond willow	same as almond willow
Willow, grey	Salix cinerea	T W	Cu	C	same as almond willow	same as almond willow
Willow, white	Salix alba	T W	Cu	C Po	same as almond willow	same as almond willow
Yew	Taxus baccata	T N-Alk	S Cu	Hf	furniture, steam bending carving	edible flesh on berry (seed within berry very poisonous)

TREES FOR DIFFERENT CONDITIONS

Appendix Two

TREES FOR COASTAL SITUATIONS

Acacia decurrens var. dealbata
Acer campestre*
Acer pseudoplatanus*+
Alnus glutinosa*
Ailanthus altissima
Amelanchier lamarckii
Araucaria araucana
Betula pendula*
Cornus sanguinea*
Crataegus monogyna*
Cupressus macrocarpa+
Eleagnus angustifolia*
Escallonia macrantha+
Fraxinus excelsior*

Hippophae rhamnoides+
Ilex aquifolium
Juniperus virginiana
Ligustrum lucidum
Pinus contorta*+
Pinus mugo+
Pinus muricata
Pinus nigra var. maritima*+
Pinus pinaster+
Pinus radiata
Pittosporum tenuifolium
Populus alba
Prunus serotina
Prunus spinosa*+

Quercus ilex*+
Quercus robur*
Rhamnus alaternus
Robinia pseudoacacia*
Salix alba*
Salix caprea*
Sambucus nigra*
Sorbus aria*
Sorbus aucuparia*
Tamarix gallica+
Ulex europaeus+
Ulmus angustifolia
Ulmus glabra
Ulmus stricta

TREES FOR CHALKY SOILS

Abies cephalonica
Abies cilicia
Abies pinsapo
Acer campestre*
Acer platanoides
Acer pseudoplatanus
Aesculus hippocastanum
Arbutus unedo
Buxus sempervirens
Cedrus sp.
Cercis siliquastrum
Cornus mas*
Crataegus monogyna*
Crataegus sp.

Eleagnus x ebbingei
Eleagnus maculata pungens
Euonymus europaeus*
Fagus sylvatica*
Ficus carica
Fraxinus excelsior*
Ilex aquifolium
Juglans sp.*
Juniperus communis*
Larix decidua*
Laurus nobilis
Malus sp.*
Morus nigra*
Pinus nigra

Populus alba
Prunus avium*
Prunus cerasifera
Prunus domestica
Prunus dulcis
Prunus spinosa
Pyrus sp.
Quercus ilex*
Salix sp.*
Sambucus nigra
Sorbus aria*
Tilia cordata*
Taxus baccata*
Ulmus sp.

* denotes recommended species
\+ denotes very salt tolerant (close to cliff planting)

TREES FOR CLAY SOILS

Abies sp.
Acer campestre*
Alnus sp.*
Betula sp.*
Carpinus betula*
Cornus sanguinea
Corylus avellana*

Crataegus monogyna*
Fraxinus excelsior*
Ginkgo biloba
Ilex aquifolium*
Larix decidua*
Liriodendron tulipifera
Malus sylvestris*

Prunus avium*
Prunus spinosa*
Salix caprea*
Sequoia sempervirens
Sorbus aucuparia*
Taxus baccata

TREES FOR SANDY SOILS

Acer pseudoplatanus
Alianthus altissima
Betula sp.*
Carpinus betula*
Castanea sativa*
Cercis siliquastrum
Crataegus monogyna*

Fraxinus excelsior*
Ginkgo biloba
Ilex aquifolium
Juniperus communis
Larix decidua*
Pinus sylvestris*
Quercus robur*

Quercus petraea*
Robinia pseudoacacia*
Sambucus nigra*
Sorbus aucuparia*
Tilia sp.

TREES FOR WET SOILS

Abies sp.
Acer pseudoplatanus
Acer rubrum
Alnus sp.*
Betula pubescens*
Hippophae rhamnoides
Mespilus germanica*

Populus sp.
Pyrus communis*
Salix sp.*
Taxodium distichum
Thuja occidentalis
Ulmus sp.

PLANNING FOR A SUSTAINABLE FUTURE

Appendix Three

This appendix has been extracted from a report 'Planning for Sustainable Woodlands', submitted to the Forestry Forum regarding reform of the Town and Country Planning system to implement 'A New Focus for England's Woodlands', the England Forestry Strategy.

Produced by a working party which was set up in August 1999 to put forward the views of the small woods and coppice workers sector to the Forestry Forum. The report describes current problems resulting from planning controls, and makes recommendations for reform, it was published in May 2000.

The working party comprised:
Lucy Nichol, PhD student, School of Planning, Oxford Brookes University.
Simon Fairlie, forest worker, author of Low Impact Development and founder of Chapter 7.
Ben Law, coppice worker, West Sussex.
Russell Rowley, chief executive, National Small Woods Association.

The following organisations support the broad aims of this document and urge the government to consider its recommendations:

- The National Small Woods Association
- Chapter Seven
- The Woodland Trust
- The Marches Greenwood Network
- Forestry Contracting Association
- British Horse Loggers Specialist Group
- British Charcoal and Coppice Group
- The Bioregional Charcoal Company
- Association of Pole-Lathe Turners
- The New Woodmanship Trust
- Common Ground
- Berks, Bucks and Oxon Greenwood Network
- Green Light Trust
- Tree Spirit
- SEEDS: Slough Environmental Education Development Service
- Surrey and Sussex Coppice Group
- Dorset Coppice Group
- Coppice Association North West
- Kentish Cobnuts Association
- Clun Valley Alder Charcoal Project
- Greater Exmoor Initiative
- Clissett Trust

INTRODUCTION

Planning controls over forestry are out of date and do not reflect the vision of sustainable multipurpose forestry now held by the government and its agencies.

Planning policies are needed which integrate economic and social aims with conservation objectives for wildlife, landscape and environment, to deliver woodland that enhances local distinctiveness and promotes a sustainable social and economic fabric on the land.

ISSUES CONCERNING PLANNING & SUSTAINABLE WOODLAND MANAGEMENT

There are three main areas for planning policy reform:

1. Definition Of Forestry

Forestry is not defined in planning law. The Town and Country Planning Act 1990 section 55 states that 'the use of any land for the purposes of agriculture or forestry (including afforestation) and the use for any of those purposes of any building occupied together with land so used' is not categorised as development, and is outside planning control. Section 336 of the same Act defines agriculture by giving a list of activities which are included. However, there is no corresponding definition of forestry, and this is leading to confusion about the scope of activities that need planning permission. For example, a local planner at a recent appeal argued that forestry was just 'the growing of a utilisable crop'. This very narrow approach would rule out even such basic forestry operations as felling. A broad definition bringing in planting, management, felling, small-scale low intensity processing and retailing of timber and timber products would remove doubt and assist in the sustainable management of woodlands.

The England Forestry Strategy calls for multipurpose woodlands. In the foreword, Elliot Morley calls for 'a great variety of well-managed woodlands' including 'woodlands for timber production to strengthen local economies' and woods for recreational use and biodiversity protection. Forestry is no longer just about single purpose plantations, and this should be reflected in planning law and policies.

Recommendation
That a definition of forestry be devised, taking in a broad spectrum of woodland uses, forestry operations and small-scale processing. The definition in the Charter of the Institute of Chartered Foresters could provide a useful starting point: 'forestry shall include all aspects of the science, economics, conservation,

amenity and art of establishing, cultivating, protecting, managing, harvesting and marketing of forests, woodlands, trees, timber and wood'.

2. Ancillary Uses

Planning controls over agriculture allow a range of processing activities of farm produce as ancillary and therefore requiring no planning permission. However, the situation for forestry operations is unclear. There are a number of products which working woodlands can provide, such as sawn timber, fencing, firewood, biomass energy, charcoal, furniture, baskets, thatching materials, brushwood, faggots, chestnuts, forest honey, saps, fibres and fruit. However, one appeal inspector stated that the only activities ancillary to forestry were the rough processing of planks and poles.

There does not seem to be any agreement about whether on-site processing, charcoal burning, etc. are ancillary uses (which would not require planning permission) or separate industrial uses (which would). Hence the discrepancies in appeal decisions. There needs to be clarification and firm support for value adding in England's woods.

The objective of establishing a sustainable and diverse forestry sector has been accepted at both national and international policy levels. Internationally the 'Agenda 21' document advocates 'promoting small-scale forest-based enterprise for supporting rural development and local entrepreneurship'. Nationally, the United Kingdom Woodland Assurance Scheme (UKWAS), agreed in June 1999, enables UK wood producers to certify their products as environmentally friendly, and use the Forestry Stewardship Council (FSC) logo. The UKWAS was designed to comply with the international standards in Principles and Criteria for Forest Stewardship laid down by the Forest Stewardship Council. These principles include local processing of the forest's diversity of products to strengthen and diversify the local economy, and avoiding dependence on a single forest product (principle 5). This makes it essential that regulations like planning are assessed to test whether they are acting as barriers to those trying to deliver sustainable forest management as defined by the Forest Stewardship Council and adopted by the government.

It is not only the owners and managers of large forest estates who will deliver the vision of FSC accredited sustainable forestry for the UK. Around half the broadleafed woodlands in Britain are small (under 25 acres/10 hectares). Of these there are approximately 430,000 acres (175,000 hectares) of derelict woods which have had little or no management for over 30 years (National Small Woods Association). A working woodland provides benefits for wildlife as well as livelihoods for rural workers. 'All woods need management to thrive and it is the working woodland that has the best chance of survival' (National Small Woods Association). Woodlands do not have to be large to support an income; even small woods can be profitable if products are processed and value is added. Wood can be sorted and sold as firewood and other wood fuel such as kindling, woodchip or compressed brash. It can be made into furniture or charcoal, restoring traditional skills such as charcoal burning and the use of pole lathes. Food stuffs such as berries, chestnuts and forest honey can be harvested seasonally from woodlands, and year-round income can be provided from cultivating edible fungi such as shiitake, which can generate considerable income throughout the year from a small area of woodland.

However, in planning, it is unclear whether these uses are ancillary to forestry or industrial processes. If they are industrial then all charcoal burners should have planning permission, which they do not and never have had. If they are ancillary uses then sheds and workshops in which to carry out these ancillary forest practices should be permitted development.

The situation is further complicated by the fact that nowadays woodland is regarded as a recreational and educational asset for the nation. Woodland owners are encouraged to open their woods to the public and some grants are only given on condition that they do so. Recreational and educational pursuits are increasingly becoming an integral part of woodland management regimes. According to Tony Phillips, Chairman of the National Small Woods Association, 'recreation and educational use are incidental to woodlands, and the generation of short-term income from such activities is one of the mainstays to underwrite the long-term growing of timber crops', (in the case of rearing game for shooting, this has been the case for many years). Such activities are analogous to the farm diversification outlined in paragraphs 1.7 and 3.4 of PPG7. However they are often particularly crucial for woodland management because the crop cycle is so long-term, and because pressures for public access to woodland are greater than for public access to cropped farmland.

Recommendations

This issue could be cleared up by providing a broad ranging definition of forestry and/or by amending the General Permitted Development Order Schedule 2, Part 7 (Forestry) so that specified activities of low environmental impact are identified as 'permitted development'. This would have the advantages of being specific about the activities and normally allowing them to take place without the need for planning permission, but also allowing local planning authorities to require the submission of planning applications through the making of Article 4 Directions, removing the permitted development rights where appropriate, or through the extension of the prior notification procedure.

Planning Policy Guidance could be amended to include words of support for value adding, local wood products and the provision of appropriate educational and recreational facilities, so that if planning permission were to be required, Local Authorities would be guided to approve it. This would bring forestry in line with agriculture by mirroring the support that exists for local food in PPG7, 1997, paragraph 3.4.

Specifically the wording of PPG7 Annex C paragraph 23 could be expanded to:

"To help ensure the long-term sustainability of small woodlands, the government wishes to maintain and develop markets for woodland produce and to encourage woodland-based enterprise that adds to rural diversification. Small-scale processing and woodland craft activities (such as charcoal burning, hurdle making and small-scale sawmills), where they are reasonably necessary to make the product marketable or disposable for profit and where they are consequential on the operations involved in producing timber grown upon the forestry unit, should be regarded as ancillary to the forestry activity and therefore do not require specific planning permission. Woodland can be particularly suitable for commercial recreation, catering for numbers of people and activities that might be intrusive in open countryside, and for educational pursuits. Such activities may sometimes play a significant role in underwriting the economics of sustainable woodland management and timber production. The Forestry Commission can advise local planning authorities on woodland recreation and education."

Listing the activities to be permitted as ancillary in this way would remove any danger of undesirable activities being conducted in woodlands under this waiver, for example producing coal tar oil, or building roads, or stacking large quantities of timber.

3. Forestry Dwellings

Up until earlier this century it was usual for bodgers and charcoal burners to live on site in woodlands. Only in the last few decades has this lifestyle died out. Modern forestry management systems, based on monocropping and clearfelling, tend to employ a peripatetic workforce consisting of a small number of highly mechanised contractors.

However the pendulum is now swinging the other way. Ecologically sustainable woodland management requires a different style of management from that adopted by modern commercial forestry. The emphasis is returning to a more varied harvest of timber, forest products and crafts, with greater employment opportunities, whilst monitoring and maintaining wildlife and biodiversity have become a priority. Continuous cover systems are now preferred to clearfell monocultures, and continuous cover requires continuous human attention. Sustainable woodland management involves a closer and more permanent relationship with the wood, which in turn suggests a greater need for on-site dwellings. These dwellings need not be of a design that diminishes the beauty of a woodland, but can be ecological building designs made of natural materials that blend into the woodland and have a low environmental impact.

There are a number of practical reasons why some woodland workers need to live on site. When making charcoal in a kiln, each burn can last up to 36 hours. Charcoal burners need to check and adjust the ventilation of their kilns, opening and closing different chimneys depending on the colour of the smoke, in the same way shepherds have to check on their ewes during the lambing season. As the burn lasts throughout the night it is not reasonable to expect a burner to travel in at 3a.m. from the village. At the end of the burn the air supply is shut off with soil and the timing is critical to ensure fully carbonised charcoal. Ensuring the safety of members of the public around the kiln is also of paramount importance. Public access in a working woodland is best managed by a resident forester, who can monitor equipment and operations out of hours. Tree nurseries require year-round protection, and human presence on a woodland site has been shown to be beneficial in deterring deer grazing on newly planted trees and regenerating coppice.

Beyond any singular functional reason lies a more general imperative. The keeper's cottage was traditionally sited within, or at the least on the very edge of the woodland. Today's woodland stewardship is perhaps more likely to be concerned with conserving biodiversity than with conserving game, but the requirements are similar. To do the job to the highest standards requires an intimate knowledge and regular surveillance of what is happening in the wood. And a lot of what happens in woods, happens at night. For instance the night-time call of the nightjar and song of the nightingale announce their arrival for the summer season. Being in the woods at night, woodland workers can pick up this information and adjust their work patterns to avoid disturbing the areas in which these birds are likely to be nesting.

There are also important economic benefits to be obtained from residence on site which can help to underwrite the economics of an otherwise marginal enterprise. Woods provide goods in kind to those living in them such as free firewood, building materials and seasonal foods. Planning guidance (PPG7, 1997 Annex I, paragraph 1) encourages rural workers to live in villages and commute to their place of work. This immediately places enormous costs on the worker, such as rent (which is often expensive and in short supply in desirable villages) and the need for a car (since public transport in rural areas is often poor and unlikely to lead to the place of work). These expenses can be avoided by those living at their place of work. The juxtaposition of employment, retail and residential uses is advocated in PPG13 which promotes the mixing of uses in order to reduce the need for travel.

The General Permitted Development Order 1995 Part 5 allows forestry workers to live in woods on a seasonal basis in caravans. It does not specify which seasons are permitted. The winter months up until the end of April are the main season for coppicing and planting; hazel work, such as hurdle making, is done around April; oak bark peeling is done in May. New plantings and fresh coppice stools need to be kept free from weeds throughout the growing season. However, conifers can be felled at any time of year, and management of rural rides, timber processing and craft activities require work all year round. In short, diverse sustainable woodland management does not have set seasons and permanent on-site accommodation would be beneficial.

PPG7, 1997, Annex I, paragraph 17, states that under conventional forestry methods a dwelling is unlikely to be justified, which presumably indicates that unconventional forestry enterprises may justify one. However applications and appeals for dwellings associated with forestry have continued to be turned down. A clear statement is needed to the effect that sustainable forestry management may require permanent on-site accommodation.

Since the late 1960s, the Government has advised Local Planning Authorities to assess whether applications for new agricultural dwellings are justified on the grounds of agricultural viability. Initially the policies were simple and flexible, for example 'viability in this context can for practical purposes be defined as offering a competent farmer the prospect of a sufficient livelihood' (Department Of Environment Circular 24/73). However, in January 1992 a revised version of PPG7 was issued which contained new policies with the aim of lessening the abuse of planning concessions for agricultural and forestry dwellings. Since then planners have applied rigorous tests to applications for agriculture and forestry dwellings to assess whether the business is financially viable and whether the dwelling is functionally necessary. However, these tests were designed for conventional methods of agriculture and forestry and are inappropriate for those wishing to live on-site and practise sustainable forest management (or indeed associated activities such as permaculture or forest gardening). Guidance on both tests needs to be adjusted, but we recognise that this can only take place in the context of a revised approach, not only to the needs of sustainable forestry, but also to sustainable agriculture as a whole. The recommendations made in the following three paragraphs are therefore long-term and with implications which reach beyond the remit of the Forum, but we feel it is important to raise them.

The functional test needs to be modified so that it does not simply refer to one isolated functional requirement (such as attending to animals or surveillance of charcoal kilns), but so that it takes into account the sum total of benefits that might accrue from living on-site. If on-site residence significantly improves the sustainable management and economic viability of the site holding (and one factor in this equation will be the amount of time and petrol spent commuting to and from the site) then this should be regarded as a material consideration in its favour.

The financial test needs to be modified to reflect the recent decision in the Court of Appeal (Arthur Sidney Petter and Monica Mary Harris v. SoS and Chichester DC, 1999) that 'The financial test is only relevant in the determination of whether the grant of permission, in whatever terms it may be granted, would, because of the uncertain future of the agricultural activity, threaten to produce in the future a non-conforming residential use that would pass with the land a use that had lost its agricultural justification'. Under certain modes of agricultural production, the judges viewed that 'profitability was no guide to the genuineness and a poor guide to probable continuation'. In the case of Mr Petter, his subsistence-based 'unit was sustainable in his hands and in that sense viable and likely to continue so'. A significant intermediary step in this direction could be made by taking into account, in any valuation of the enterprise, the value of goods in kind at the retail price that they would otherwise have to be acquired, rather than at 'farm-gate' or 'roadside' price.

Within the context of sustainable land management, the financial and functional tests, as they are phrased at the moment, are very blunt instruments. To safeguard against abuse and prevent the proliferation of rural dwellings for those not working on land, a more sophisticated criteria-based approach could be used. This would ensure that only those engaged in genuine sustainable land-based activities could benefit from it. It would also guarantee that sustainable operations could never lapse into unsustainable ones (for example through a change in ownership). The mechanisms are already available in planning law in the form of planning obligations (Section 106 agreements), but have yet to be used to their full potential by local authorities. By stating a list of criteria in a local plan ('development will be permitted provided that') a local authority could use this as the basis for a conditional planning approval, with a Section 106 legal agreement ensuring that the criteria be met. Model criteria have been prepared by Chapter Seven in their document 'Defining Rural Sustainability: Fifteen Criteria for Sustainable Development in the Countryside' (1999, see Annex I later in this document). The use of criteria in planning has been endorsed by the Countryside Agency in Planning for Quality of Life in Rural England, the interim planning policy of the Countryside Agency (1999). This takes forward the work on Countryside Character developed by the former Countryside Commission, and recognises that criteria provide a method for delivering developments of a specified type while blocking those developments deemed undesirable.

Recommendations

That paragraph 16 in Annex I of PPG7 is amended to read:

"Local Planning Authorities should normally apply the same criteria to applications for forestry dwellings as to agricultural dwellings. Under conventional modern methods of forestry management, which use a largely peripatetic workforce, a new forestry dwelling is unlikely to be justified. However, current forestry policy recognises that there is a large area of neglected woodland in England which requires management and which could contribute to the local rural economy. In special circumstances where it is necessary for the sustainable management of woodlands, residential development should be permitted. Initially this would be only on a seasonal or temporary basis. Residential structures should be of a siting, scale and design appropriate to the surroundings and the scale of the operation; and environmentally sound management of the woodland should be secured by condition or legal agreement."

Also recommended are that policies are written into local plans stating that on-site accommodation for forestry workers will be permitted subject to adherence to certain criteria which would be enshrined in conditions or a Section 106 agreement. For example, the woodland worker would be legally committed to engage in sustainable woodland management, use renewable energy, restrict their use of fossil fuel powered vehicles, and build the house and workshops from local natural materials. Additional criteria should apply in the case of ancient woodlands to ensure that the location of a dwelling and the access to it would not have a negative impact on the integrity of the site.

Also recommended is that the value of goods in kind provided by the wood are discounted in any calculation of financial viability at the full retail value (which is what they would cost if bought in from outside) rather than at 'roadside' or 'farm-gate' prices. This is important, because the Agricultural Development Advisory Service (ADAS) do discount subsistence production, but at production prices rather than retail prices. This change could be explained in a Good Practice Guide, and ADAS staff and planners could receive advice and training on the sustainable rural livelihoods model.

Update General Permitted Development Order (GPDO) Part 5 on caravans to include 'wooden structure which can be easily dismantled' in line with PPG7, 1997, Annex I, Part 14. This would give seasonal workers the option of living in low impact wooden self-built structures; modern equivalents of the simple dwellings built by bodgers and charcoal burners in the past. Since these homes are made from materials sourced from the woods themselves, they are more appropriate to their setting than a conventional caravan and revive a vernacular British housing tradition. They also have a very low embodied energy (the energy used in the manufacture, transport, use and disposal of a commodity), so are extremely environmentally friendly.

ANNEX 1

TWO EXTRACTS FROM *DEFINING RURAL SUSTAINABILITY* CHAPTER 7 (1999)

Full report available from Chapter 7, The Potato Store, Flax Drayton Farm, South Petherton, Somerset TA13.

Fifteen Criteria for Developments

Associated With Sustainable Land-Based Rural Activities.

1. The project has a management plan which demonstrates: how the site will contribute significantly towards the occupiers' livelihoods; how the objectives cited in items 2 to 14 below will be achieved and maintained.

2. The project provides affordable access to land and/or housing to people in need.

3. The project provides public access to the countryside, including temporary access such as open days and educational visits.

4. The project can demonstrate how it will be integrated into the local economy and community.

5. The project can demonstrate that no activities pursued on the site shall cause undue nuisance to neighbours or the public.

6. The project has prepared a strategy for the minimisation of motor vehicle use.

7. The development and any buildings associated with it are appropriately sited in relation to local landscape, natural resources and settlement patterns.

8. New buildings and dwellings are not visually intrusive nor of a scale disproportionate to the site and the scale of the operation; are constructed from materials with low embodied energy and environmental impact, and preferably from locally sourced materials, unless environmental considerations or the use of reclaimed materials determine otherwise. Reuse and conversion of existing buildings on the site is carried out as far as practicable in conformity with these criteria.

9. The project is reversible, insofar as new buildings can be easily dismantled and the land easily restored to its former condition.

10. The project has a strategy to minimise the creation of waste and to reuse and recycle as much as possible on-site.

11. The project has a strategy for energy conservation, and the reduction over time, of dependence on non-renewable energy sources to a practical minimum.

12. The project aims over time for the autonomous provision of water, energy and sewage disposal and where it is not already connected to the utilities, shall make no demands upon the existing infrastructure.

13. Agricultural, forestry and similar land-based activities are carried out according to sustainable principles. Preference will be given to projects which conform to registered organic standards, sustainable forestry standards or recognised permaculture principles.

14. The project has strategies and programmes for the ecological management of the site, including: the sustainable management and improvement of soil structure; the conservation and, where appropriate, the enhancement of semi-natural habitat, taking into account biodiversity, indigenous species, and wildlife corridors; the efficient use and reuse of water, as well as increasing the water holding capacity of the site; the planting of trees and hedges, particularly in areas where the tree coverage is less than 20%.

15. The project can show that affordability and sustainability are secured, for example, by the involvement of a housing association, co-operative, trust or other social body whose continuing interest in the property will ensure control over subsequent changes of ownership.

THREE MODEL POLICIES FOR LOCAL PLANS

Model Policy A: *Sustainable Land-based Economic Activities*

Developments associated with agriculture, forestry and other land-based economic activities in the open countryside should be sustainable and should demonstrate that they, and the activities concerned, have a beneficial or minimally adverse impact on the surrounding countryside and its occupants, and on the wider environment. In judging the sustainability of such activities, regard will be had to:

Ecological management of the site.

Farming and forestry methods.

Waste, energy and resource management.

The siting and structure of buildings.

Vehicle use.

The impact on the surrounding community.

Public access.

Model Policy B:
Dwellings Associated with Sustainable Land-Based Activities
Applications for isolated dwellings in the countryside will be permitted where:

It is essential for the proper functioning of the enterprise for one or more workers to be readily available at most times.

There is clear evidence that the proposed enterprise has been planned on a sound financial basis and with a firm intention and ability to carry it out.

On-site residence will help to minimise, rather than increase, overall car use.

The dwelling and its services will, as far as is practicable, be sustainably designed in regard to siting, scale, resource use and environmental impact.

Other planning policies, for example on access or design, are satisfied. Dwellings associated with new sustainable land-based enterprises will initially only be granted planning permission for a temporary period of three years.

Model Policy C:
Sustainable Affordable Housing
In considering new dwellings associated with sustainable land-based activities, the council may exceptionally grant permission for a sustainable housing development where the need for land-based occupancy is part-time, seasonal or inconclusive, provided that:

Occupancy is restricted by legal agreement or condition to those meeting the criteria of local need for affordable housing.

The dwelling and its services will, as far as is practicable, be sustainably designed in regard to siting, scale, resource use and environmental impact.

The scale and nature of the dwellings meet the criteria for affordable housing identified in this plan.

Full copies of the document, *Planning for Sustainable Woodlands* available from: Chapter 7:
www.tlio.org.uk/chapter7

Appendix Four

RELEVANT FORESTRY PLANNING CASES

Ben Law
Chichester District Council
6 April 1998
Temporary planning permission for forestry dwelling and forestry workshop.
Application No.: LD/97/02695/FUL and LD/97/02783/FUL

Ben Law
Chichester District Council
4 June 2001
Full planning permission for chestnut framed, straw bale walled forestry dwelling. Limited to Mr Law personally and to be removed if he ceases working the land.
Application No.: LD/01/00296/FUL

Ben Law
Chichester District Council
6 June 2012
Removal of personal tie and demolition of building. Occupation of dwelling tied to forestry worker at Prickly Nut Wood.
Application No.: LD/11/05125/FULNP

Hugh Ross and Caroline Church
Kettering Borough Council
22 April 1996
Temporary planning permission for forestry dwelling.
Application No.: KE/96/0242 Renewed 1999.

Hugh Ross and Caroline Church
Kettering Borough Council
30 April 2004
Full planning for timber framed forestry dwelling.
Application No.: KE/04/0465

Middlemarsh
West Dorset District Council
Refusal of an application for a forestry dwelling.
Application No.: T/APP/F1230/A/99/1022133/P4

National Trust
Pembrokeshire Coast National Park
13 March 2000
Temporary planning permission for up to ten low impact dwellings for one year for forestry use.
Application No.: NP/00/117

Chris Wall-Palmer
Chichester District Council
2 April 2008
Full planning permission for forestry dwelling with personal tie.
Application No.: SJ/08/00591/FUL
Personal tie removed on 13 May 2008, dwelling tied to forestry use with charcoal burning.

Tinkers Bubble
South Somerset District Council
7 January 1999
Temporary planning permission for a group of people, mixed agriculture/forestry enterprise.
Application No.: 980/1651

R. Waterfield
Denbighshire County Council
Appeal against planning contravention notice in respect of charcoal burning being interpreted as industrial rather than forestry use.
Application No.: APP/R6830/A/99/512773/T

SUSTAINABLE RURAL LIVELIHOODS

Appendix Five

DEPARTMENT FOR INTERNATIONAL DEVELOPMENT'S DEFINITIONS

Financial capital
The financial resources which are available to people (whether savings, supplies of credit or regular remittances or pensions) and which provide them with different livelihood options.

Human capital
The skills, knowledge, ability to labour and good health important to the ability to pursue different livelihood strategies.

Natural capital
The natural resource stocks from which resource flows useful for livelihoods are derived (e.g. water, wildlife, biodiversity, environmental resources).

Physical capital
The basic infrastructure (transport, shelter, water, energy and communications) and the production equipment and means which enable people to pursue their livelihoods.

Social capital
The social resources (networks, membership of groups, relationships of trust, access to wider institutions of society) upon which people draw in pursuit of livelihoods.

Appendix Six — EXAMPLE OF CONTRACT

Contract for cutting coppice woodland between .... (hereinafter referred to as the land owner) and .... (hereinafter referred to as the coppice worker).

1. The coppice woodland to be cut is called .... and the boundaries of the agreed area to be cut are outlined in red on the enclosed map, scale 1:2500.

2. It is the responsibility of the land owner to ensure that the woodland has either a felling licence or is under a Woodland Grant Scheme contract unless the coppice is less than 6in (15cm) in diameter, 4ft3in (1.3m) from the ground. Copies of any Forestry Commission contract affecting the woodland should be given to the coppice worker.

3. The period of this contract for cutting coppice commences on .... until ....

4. The period of this contract for adding value in the woodland to the cut coppice commences .... until .... If adding value is taking place in the woodland, the coppice worker should list the type of adding value work to be undertaken ....

5. The routes for extraction of produce from the woodland should be marked on the enclosed map outlined in blue. The period of this contract for extraction of produce commences .... until................

6. The period of this contract for the coppice worker to reside on the land while cutting and or adding value under the Caravan Site and Control of Development Act 1960 as a seasonal forestry worker commences .... until ....

7. Deer and rabbit control, delete as necessary:

a) The land owner will take full responsibility for deer and rabbit control in the woodland after coppicing.

b) The coppice worker will construct brash hedges around the cut coppice of .... in height and .... in width, all other responsibility for deer and rabbit control is that of the land owner.

c) A deer fence of .... in height and a rabbit fence of................ in height, dug into the ground .... inches will be constructed by .... all other responsibility for deer and rabbit control is that of the land owner.

d) Other suggestions ....

8. The coppice worker shall have a maximum of .... burning sites per acre.

9. The condition of the extraction rides at the beginning of this contract is .... The coppice worker shall ensure that he leaves the rides in the same condition as found unless specified to improve them by the land owner. If the land owner wishes for the rides to be improved this should be laid out under point 13.

10. At the termination of this contract the coppice worker agrees to leave the woodland, tidy of piles of brash unless otherwise specified by the land owner and clean of rubbish. Provided this is done to the satisfaction of the land owner, the land owner agrees to offer the coppice worker first refusal on the next cut.

11. Is the woodland certified under a Forest Stewardship Council Scheme? ....

12. The land owner will advise the coppice worker prior to commencement of work of any hazards or constraints associated with the woodland. The coppice worker should comply with statutory health and safety regulations, and shall have insurance cover as required by the land owner against any liability appropriate to the woodland.

13. The land owner and coppice worker agree to the following extra tasks to be undertaken within the woodland by the coppice worker during the agreed period of this contract ....

14. The financial agreement between the land owner and coppice worker is as follows .... To be paid (delete as appropriate) in instalments of .... in full by .... other arrangement ....

15. Failure as a result of an act of God, flood, fire, war, shall not be regarded as a breach of contract. If disagreement occurs the parties will seek from the parties of the Surrey and Sussex Coppice Group the services of a mediator.

Signed .... (land owner)
Date ....

Signed .... (coppice worker)
Date ....

BIBLIOGRAPHY   Appendix Seven

Titles marked † are available from:
Permanent Publications'
online retail store:
www.green-shopping.co.uk
The Sustainability Centre, East Meon,
Hampshire GU32 1HR
Tel: 01730 823 311 Fax: 01730 823 322
Email: enquiries@green-shopping.co.uk

AGROFORESTRY
The books in this section are written in studious detail by Martin Crawford and are an important addition to the forest dweller's library. All titles are published by the Agroforestry Research Trust and are available from Permanent Publications.

Agroforestry Options For Landowners†
Handy pamphlet outlining the possibilities and where to start.

Bamboos†
Covers over 100 species of bamboo for temperate climates. Includes cultivation and management, using bamboos for ground cover, hedging, cane production and edible shoot production, and extensive details of all species and varieties available in Britain.

Bee Plants†
Giving details of over 1,050 species which useful to both wild and hive bees. Details include time of year of benefit (e.g. flowering for nectar production), type of benefit (nectar, pollen, honeydew etc.), siting requirements etc.

Dye Plants†
Details over 550 species, from trees to lichens, from which dyes can be obtained and includes siting requirements, performance indicators and the colours obtained using different mordants on different fabrics.

Blackcurrants & Raspberries†
Extensive information on all aspects of cultivation of blackberries, raspberries and hybrid berries. Includes extensive cultivar lists and descriptions.

Cherries, Production & Culture†
Detailed information on cherries, including description of the many species, silviculture, cultivation and management, and an extensive cultivar list with descriptions.

Currants & Gooseberries†
Blackcurrants, red and whitecurrants and gooseberries are all covered extensively. Descriptions and uses are given, along with cultivation details and cultivar lists and descriptions.

Directory of Apple Cultivars†
Describes over 3,000 varieties and covers choosing cultivars for any conditions.

Directory of Pear Cultivars†
Describes over 650 varieties and covers choosing appropriate cultivars for particular conditions.

Fruit Varieties Resistant to Pests & Diseases†
Lists all common garden fruits varieties (apples, pears, plums, cherries, currants, berries etc.) resistant to common pests and diseases from which they may suffer.

Ground Cover Plants†
Describes over 850 species and includes sections on grass-based ground covers, paths, and ground covers for the short, medium and long-term. Includes easy-to-read tables and sections on mixing species for better cover.

Chestnuts, Production & Culture†
Hazelnuts, Production & Culture†
Walnuts, Production & Culture†
The above three ART booklets are probably the most complete collection of information in print on growing nuts in Britain. Subjects covered include: siting, choosing cultivars for British conditions, planting, pruning, pollination, pests and diseases, harvesting and processing.

Nitrogen Fixing Plants for Temperate Climates†
Extensive coverage of nitrogen fixing plants.

Plants For Basketry†
Covers over 300 species, mainly shrubs and trees, which can be used for basketry. Details of siting and performance are given, as well as any cultural techniques normally used to provide material for basketry and other uses of species.

Plants For Hedging†
Covers some 450 species of trees and shrubs which are used for hedging and shelterbelts. Details given include siting and performance, other uses of species and any particular techniques used when being grown in hedges.

Plums, Production, Culture & Cultivar Directory†
Details how to cultivate European plums, flowering characteristics, plums recommended for drying, plums for cool regions and disease resistant cultivars. Gives a history of plums and describes each known plum species, lists the minor cultivars and gives valuable information on rootstocks.

Timber Trees For Temperate Climates[†]
A directory of timber-producing trees for temperate climates with some 500 species included. Information includes common names, origin, hardiness zones, pH, moisture and light requirements, height at 10 years of age and maximum height; timber properties (strength, durability, heaviness, hardness, shock absorbency, splitability, flexibility, fragrancy) and all reported timber uses.

Useful Plants For Temperate Climates[†]
A series of volumes detailing siting (including pH, moisture and light requirements), cultivation and UK performance indicators at 10 years (trees only) and hardiness. Lists all possible uses: culinary, medicinal, fibre, rubber, ground cover, timber, facade insulation, basketry, green manures, bee and animal fodder, etc. Over 7,500 species, from trees to lichens.
Vol 1: *Trees* (classified by size)
Vol 1a: *Trees* (alphabetical)
Vol 2: *Shrubs* (classified by size)
Vol 2a: *Shrubs* (alphabetical)
Vol 3: *Perennials*
Vol 4: *Annuals & Biennials*
Vol 5: *Algae, Fungi & Lichens*
Vol 6: *Climbers*
Vol 7: *Bulbs, Roots & Tubers*

GENERAL

Ancient Woodland[†]
Oliver Rackham; Edward Arnold, 1980.
ISBN 0-71312-723-6.
The text on ancient woodland – expensive!

Building a Low Impact Roundhouse[†]
Tony Wrench; Permanent Publications, 2001.

Caring For Small Woods
Ken Broad; Earthscan, 1998.
Practical advice from an experienced forester.

Collins Guide To Tree Planting & Their Cultivation
H. Edlin; Collins, 1975.
ISBN 0-00212-002-X
Recommended guide. Out of print.

Coppicing and Coppice Crafts[†]
Rebecca Oaks and Edward Mills; Crowood Press, 2010.
Good introduction to coppicing and crafts with Cumbrian flavour.

Creating New Native Woodlands[†]
J. Rodwell & G. Patterson; Forestry Authority Bulletin No.112, 1994.
Useful for species classifications.

Digest 445, Advance In Timber Grading
BRE Centre For Timber Technology and Construction
Tel: 020 7505 6622
Reference for chestnut receiving a British Standard within the building industry.

EcoForestry[†]
Alan Drengson & Duncan Taylor; New Society, 1997.
Collection of articles – an important read.

*Earth Care Manual:
A Permaculture Handbook For Britain & Other Temperate Climates*[†]
Patrick Whitefield, Permanent Publications, 2nd edition 2011.
A comprehensive explanation of permaculture design for cool climates.

Flora Brittanica
Richard Mabey; Trafalgar Square, 1996.
The evolving culture of our flora – excellent.

Handbook For Croft Forestry
Bernard Planterose, 1992.
Practical guide for upland situations.

Hedging
Alan Brooks; BTCV, 1988.
Good practical section on regional variations.

History Of The Countryside
Oliver Rackham; Phoenix Press.
A fascinating description of how the British landscape and human activities have interacted over many centuries to create what we see today.

Managing Irregular Forests
Association Futaie Irreguliere, 2010.
French approach to continuous cover forestry.

Orchards – A Guide To Local Conservation
Common Ground, 1989.
Cultural identity of our orchard heritage.

Permaculture: Principles & Pathways[†]
David Holmgren; Permanent Publications, 2002.
The classic update on permaculture design and energy descent.

Practical Forestry For The Agent & Surveyor
C. Hart; Alan Sutton, 1991.
Used by forestry students.

Reclaiming Degraded Land For Forestry
Moffat & McNeill; Forestry Commission Bulletin No.110, 1994.

Roundwood Small-diameter Timber for Construction
VTT Publications 383, 1999.
ISBN 951-38-5387-X
Structural testing of Roundwood.

Roundwood Timber Framing[†]
Ben Law; Permanent Publications, 2010.
The how to build with roundwood book.

Silviculture Of Broadleaved Woodland
J. Evans; Forestry Commission Bulletin No.64, 1984.
Highly recommended text.

Trees & Aftercare[†]
Barbara Kiser; BTCV.
Practical, well illustrated.

Trees & Woodlands In The British Landscape
Oliver Rackham; Weidenfeld & Nicolson, 1995.
ISBN 0-46004-786-8
Excellent historical perspective.

Woodland Conservation & Management
G.F. Peterken; Chapman & Hall, 1981.
Nature conservation perspective.

Woodlands – A Practical Handbook
Alan Brooks; BTCV, 1992.
Practical, well illustrated.

Woodland House[†]
Ben Law; Permanent Publications, 2005.
Follows the building of the author's sweet chestnut home in his woodland, Prickly Nut Wood.

Woodland Management[†]
Chris Starr; Crowood Press, 2005.
Well rounded woodland management book.

Woodland Year[†]
Ben Law; Permanent Publications, 2008.
Author's month by month guide to working his woods.

PERIODICALS
Agroforestry News
www.agroforestry.co.uk

The Dendrologist
www.dendrologist.org.uk

Fine Woodworking
www.finewoodworking.org

Forestry Journal
www.forestryjournal.co.uk

The Land Magazine
www.thelandmagazine.org.uk

Living Woods Magazine
www.living-woods.com

Permaculture Magazine[†]
www.permaculture.co.uk

Reforesting Scotland
www.reforestingscotland.org

Tree News
www.treecouncil.org.uk

Woodlots
www.woodnet.org.uk

PLANNING & LAND ISSUES
Low Impact Development[†]
Simon Fairlie; Jon Carpenter, 1996.
The only enjoyable read on planning and rural development!

This Land Is Our Land
M. Shoard; Gaia Books, 1997.
Well researched read on land ownership and history – read it!

The Living Land
J. Pretty; Earthscan, 1998.
A positive message about the regeneration of the countryside – a clear case for sustainable agriculture.

Sustainable Rural Livelihoods
D. Carney; Department for International Development, 1998.

WOODLAND CRAFTS
Wood & How To Dry It[†]
A fine Woodworking Publication; Taunton Press, 1986.
Excellent collection of wood drying information with practical designs.

Going With The Grain
Mike Abbott; Living Woods Books, 2011.
Detailed chairmaking book from tree to finished product from one of the country's most experienced chairmaking tutors.

Greenwood Crafts[†]
Edward Mills and Rebecca Oaks: Crowood Press, 2012.
Excellent compliment to their coppicing and coppice crafts book.

Living Woods (Revisited)
Mike Abbott; Living Woods Books, 2013.
Mike's journey of working in the wood.

Making Rustic Furniture
Daniel Mack; Lark, 1990.
Inspirational.

The Rustic Furniture Companion
Daniel Mack; Lark, 1996.
Further inspiration.

Tools & Devices For Coppice Crafts
F. Lambert; Centre for Alternative Technology, first published 1957.
Hard to read, but many useful designs.

Traditional Woodland Crafts
Raymond Tabor; B.T. Batsford, 1994.
Well illustrated and practical.

Woodcolliers & Charcoal Burners
Lyn Armstrong; Coach Publishing House Ltd. and the Weald and Downland Open Air Museum, 1978.
Historical perspective – not practical.

Woodland Crafts In Britain
H. Edlin; David & Charles, 1973.
ISBN 0-71535-852-9
Classic text. Out of print.

WOODLAND FOOD
Creating a Forest Garden[†]
Martin Crawford; Green Books, 2010.
Excellent species book.

Food for Free[†]
Richard Mabey; Collins, 1989.
The classic pocket guide to wild food.

How To Make A Forest Garden[†]
Patrick Whitefield; Permanent Publications, 1996.
Gives good design information.

Mushrooms & Other Fungi Of Great Britain & Europe[†]
Roger Phillips; Macmillan, 1994.
Well photographed guide book.

Mycelium Running[†]
Paul Stamets; Ten Speed Press, 1994.
Fascinating look into the world of mushrooms.

Plants For A Future[†]
Ken Fern; Permanent Publications, 1997.
An opening door to the vast possibilities of edible and useful temperate climate species.

Wild Food[†]
Roger Phillips; Pan Books, 1983.
Beautifully photographed with good recipes.

Some books selected by the author in this Bibliography are out of print. It is sometimes possible, however, to track down secondhand copies. International Standard Book Numbers (ISBN) have therefore been given with books listed here as out of print in the hope that this will help you find them.

RESOURCES   Appendix Eight

COMMUNITY FORESTS

Central Scotland Forest
Lanarkshire
www.csft.org.uk

Cleveland Community Forest
Middlesbrough, Cleveland
www.teesforest.org.uk

Forest of Avon, Bristol
www.forestofavon.org.uk

Forest of Mercia
Staffordshire
www.forestofmercia.org.uk

Great North Forest
County Durham
www.communityforest.org.uk

Great Western Community Forest
Swindon
www.swindon.gov.uk/forest/

Greenwood Forest
Nottingham
www.greenwoodforest.org.uk

Thames Chase, Essex
www.thameschase.org.uk

Marston Vale, Bedfordshire
www.marstonvale.org.uk

Mersey Forest, Warrington
www.merseyforest.org.uk

Red Rose Forest, Manchester
www.redroseforest.co.uk

South Yorkshire Forest, Sheffield
www.syforest.co.uk

Watling Chase, Hertfordshire
www.watlingchase.org.uk

COPPICE GROUPS

Coppice Director
www.coppice-products.co.uk

Coppice Association North West (CANW)
www.coppicenorthwest.org.uk

Dorset Coppice Group
www.dorsetcoppicegroup.co.uk

Hampshire Coppice Group
www.hampshirecoppice.co.uk

Malvern Coppice Network
www.malvernhillscoppicenetwork.org.uk

Sussex & Surrey Coppice Group
www.coppicegroup.org.uk
www.coppicegroup.wordpress.com

Three Ridings Coppice Group
www.three-ridings.org.uk

MOBILE SAWMILLS

Logosol
www.logosol.co.uk
LumberMate and chainsawmills

LogLogic
www.loglogic.co.uk
Autotrek large capacity British designed and built

Lucas Mill
www.lucasmill.com

Mahoe Mills NZ
www.mobilesawmills.co.uk

Pezzalotto Sawmills
www.sawsuk.com

Wood-mizer UK
www.woodmizer.co.uk

MUSHROOM SPAWN SUPPLIERS

Anne Miller
www.annforfungi.co.uk

Humungus Fungus
www.humungusfungus.co.uk

TREE NURSERIES

BTCV Trees and Shrubs
www.tree-life-centre@tcv.org.uk

Butterworths Organic Nursery
Garden Cottage, Auchinleck, Cumnock,
Ayrshire KA18 2LR
Tel: 01290 551 088
www.butterworthsorganicnursery.co.uk

Castle Howard Forest Nursery
Castle Howard, York YO60 7DA
www.castlehoward.co.uk

Cheviot Trees
Newton Brae, Foulden, Berwick-upon-Tweed, Scotland
www.cheviot-trees.co.uk

Chew Valley Trees
Winford Road, Chew Magna,
Bristol BS18 8QE
www.chewvalleytrees.co.uk

Cool Temperate Nursery
Cossall, Nottingham
www.cooltemperate.co.uk

Forestart
Church Farm, Hadnall,
Shrewsbury SY4 4AQ
www.forestart.co.uk

Hilliers Nursery
Ampfield House, Nr. Romsey,
Hampshire SO51 9PA
www.hilliertrees.co.uk

Nutwood Nurseries
Orsedd Bach, Llanarmon,
Pwllheli LL53 6PU
www.farcourt.co.uk/ge/nutwoods.html

Oakover Nurseries Ltd
Calehill Stables, The Leacon, Charing,
Ashford
www.oakovernurseries.co.uk

Organic Trees From Highland Seed
Doire Na Mairst, Morvern, by Oban,
Argyll PA34 5XE, Scotland
Tel: 01967 421 203
www.treenurseryscotland.org

Park Farm Nurseries
Ledbury Road, Newent, Gloucestershire
GL18 1DL
www.parkfarmforestnursery.co.uk

Spains Hall Forest Nursery
Spains Hall Farmhouse, Finchingfield,
Braintree, Essex CM7 4NJ
www.spainshall.co.uk

Willow Bank
Ragmans Lane Farm, Lower Lydbrook,
Glos. GL17 9PA
www.thewillowbank.com

Tools & Equipment

Richmonds
www.richmondsgroundcare.co.uk
Mail order supplies of all forestry equipment

Chieftain Forge Ltd
www.cheiftainforge.co.uk
Tree planting spades, shelters, etc

Horse Logging Equipment
www.heavyhorses.net
British built forwarders, timber arches etc

Plantoil UK
www.plantoil.co.uk
Biodegradeable chainsaw oil

Riko UK
www.alpinetractors.com
Suppliers of small-scale forestry extraction equipment

The Woodsmiths Store
www.woodsmithstore.co.uk
Good range of green woodworking tools, books and assessories

Woodland Craft Supplies
www.woodlandcraftsupplies.co.uk
Forged craft tools

Useful Organisations

Agricultural Mortgage Corporation
www.amconline.co.uk

Association of Pole Lathe Turners & Greenwood Workers
www.bodgers.org.uk

Basketmakers Association
www.basketassoc.org

Bioregional Development Group
www.bioregional.co.uk

British Deer Society
www.bds.org.uk

British Trust For Conservation Volunteers (BTCV)
www.btcv.org

Chapter Seven
www.tlio.org.uk/chapter7

Coed Cymru
www.coedcymru.org.uk

Common Ground
www.commonground.org.uk

Crofters' Commission
www.crofting.scotland.gov.uk

Department for Environment, Food & Rural Affairs (DEFRA)
www.defra.gsi.gov.uk

Ecology Building Society
www.ecology.co.uk

Forestry Commission
www.forestry.gov.uk

Forestry Contracting Association (FCA)
www.fcauk.com

Forest Schools
www.forestschools.com/

The Forest Stewardship Council (FSC)
www.fsc-uk.org

Friends & Families of Travellers
www.gypsy-traveller.org

Green Wood Centre
www.smallwoods.org.uk

Institute of Chartered Foresters
www.charteredforesters.org

Institute of Professional Soil Scientists
www.soilscientist.org

The Land Is Ours
www.tlio.org.uk

Lantra Awards
www.lantra-awards.co.uk

Low Impact Living Initiative
www.lowimpact.org

National Coppice Apprenticeship
www.smallwoods.org.uk

National Forest Company
www.nationalforest.org

National Proficiency Tests Council
www.nptc.org.uk

Natural England
www.naturalengland.org.uk

Permaculture Association (Britain)
www.permaculture.org.uk

Programme for the Endorsement
of Forest Certification
www.pefc.org.uk

Ramblers Association
www.ramblers.org.uk

Royal Forestry Society
www.rfs.org.uk

Royal Society for the Protection of Birds
www.rspb.org.uk

Royal Society of Wildlife Trusts
www.wildlifetrusts.org.uk

Scottish Natural Heritage
www.snh.org.uk

Sustainability Centre
www.sustainability-centre.org

The Soil Association
www.soilassociation.org

Small Woods Association
www.smallwoods.org.uk

The Tree Council
www.treecouncil.org.uk

Trees For Life
www.treesforlife.org.uk

Troidos Bank
www.triodos.co.uk

The Veteran Tree Initiative
www.naturalengland.org.uk

Woodland Heritage
www.woodlandheritage.org

The Woodland Trust
www.woodlandtrust.org.uk

Woodschool, Borders
www.woodschool.ltd.uk

The Working Horse Trust
www.theworkinghorsetrust.org

WOODLAND ESTATE AGENTS

www.johnclegg.co.uk

www.stackyard.com

www.bachellermonkhouse.com

www.boultoncooper.co.uk

www.uklandfarms.co.uk

www.forests.co.uk

www.woodlands.co.uk

www.woods4sale.co.uk

Index

Abbot, Mike 96
Acacia 66
   false 66
Acidic soil 14, 126
Acid rain 2
Adze 13
Afforestation 17, 152, 178
Agenda 21 31, 160, 179
Agriculture 147
Agrochemicals 66, 147
Agroforestry 12, 62, 63, 99
Agroforestry Research Trust 126
Air pollution 2
Albania 10, 11
Albon, Pete 95
Alder 4, 7, 9, 12, 16, 42, 46, 55, 57, 66
Algarth Insurance Brokers Limited 146
Amazon rainforest 50
Amphibians 88
Ancient woodland 2, 5, 6, 36, 42, 86, 87, 100, 182
Animal manure 49
Aquatic crops 48
Arable crops 12
Ascomycete fungus 'Frankia' 4
Ash 12, 17, 79, 108

Badger 23, 40
Barbecue charcoal 105
Baron (the horse!) 95, 96
Basket-making 108
   swill 108
Bats 88, 89
Bean poles 70
Beating up 63
Beech 7, 10, 16, 60, 62, 79, 81, 138
Bees 23, 57, 139-140
Benders 112, 153, 154
Billhook 92
Biochar 106
Biodegradation 62, 63
Biodegradeable materials 54
Biodiversity 2-3, 34, 82, 155, 178, 180, 183
Bioregional Charcoal Company 106, 107
Bioregional Development Group 106-107
Biosecurity 161

Birches 2, 14, 23, 46, 50, 52, 55, 65, 66, 83, 84, 109, 138
Bird manure 3, 4, 52
Blackberry 86
Black locust 12
Blackthorn 127
Black truffle 138
Blair, David 97-101
Bodging 107
Boletus edulis 137
   subtomentosus 137
B&Q 107
Bradfield Woods 7, 47
Bradford Hutt plan 82
Bramble 46, 50, 75, 137
Brash 71, 74, 75, 99, 102, 132, 134
Brash hedges 74, 75, 99, 102, 132, 134-137
Bridleway 36
The British Horse Loggers 95
Broadleaf plantations 14, 16, 53
Building construction 31, 84, 97
Butterflies 2, 6, 87, 89, 111
   habitat preferences of 35

Cabin 31, 84-86, 99, 112, 154
Caledonian pinewoods 6
Camphill Village Trust 118
Canopy, tree 12, 81, 124
Caravan 31, 112, 151-154
Caravan Sites and Control of Development Act (1960) 152, 154
Carbon fixing 48, 51
Celebration 161
Certificate of Competence 147
Certificate of Lawful Use 150
Certification 117-119, 158
Chainsaws 55, 84, 92, 113, 146, 147
Charcoal 2, 102, 106, 109
   burning 105-106, 146, 150
   dust 105
   production 113
Charcoal burners 21, 75, 104, 106
   sustainable rural livelihoods 21-27
Cherry Orchard Centre 107
Chestnut, sweet 37, 40, 42, 52, 101, 108, 119, 122-125, 163
   chestnut flour, analysis of 123

Marigoule 123
Marron de Lyon 123
Chichester District Council 153-154
Clay 58, 59, 66, 112, 134
Clear felling 3, 4, 80-81
Cleaving 96
Climbing plants 132
Cloud formation 4
Coal 2
Cobnuts 126
Colchester Borough Council 150
Common Agricultural Policy 147
Community forests 67, 100
Compaction 66, 94
Compost 59, 61, 100, 113, 132
Compost toilet 113, 132
Condensation 3
Conifer plantations 13, 14-15, 81, 119
Continuous cover forestry 16
Continuous Cover Forestry Groups (CCFG) 16
Contracts 71, 155
Coppiced wood 2, 10, 73, 101-104
Coppice fruit avenues
   cleft grafting (top-working) 134
   grafting 132
   planting 134
   shield budding 134
   whip and tongue grafting 132-133
Coppice woodland system 2, 6, 73-76
   buying 70-71
   establishing 70
   growth patterns 8
   management of 11, 71-73
   overstood, stored or neglected 8, 76-78
   root system 8
   short rotation 9
   with standards 7
Coppicing 2, 3, 70, 92, 105
   for wildlife 87-88
Cord wood 105
Cork 12
Corsican pine 66, 79, 99
Corylus avellana 126
   maxima 126

Countryside Agency 16, 17, 182
Countryside Council for Wales (CCW) 145
Crab apples 47, 55, 132, 134
Craftwork 52, 58, 104
Cratellus cornucopioides 137
Crayfish 89
Crofting Counties Agricultural Grant scheme 155
Crown deformation 3, 4
Crown density 2
Croydon 107
Cutting 2, 6, 11, 41, 52, 70, 82, 89, 92, 126

Dangerous Animals Act 141
Dead wood 4, 71, 73, 88
Deciduous woodland 3
Deer 2, 6, 23, 36, 46, 62, 63, 71, 73-76, 113, 140, 155
Degraded land, reclaiming 66-67
Dehesa system 11
Department for Environment, Food and Rural Affairs (DEFRA) 144, 148-149
Planning Policy Guidance Note 7 147, 148, 150-151
Department for International Development (DFID) 25, 31
Derelict Land Grant (DLG) 66
Designing woodlands
   Goudhurst 55-57
   Meera's wood 58-59
   permaculture principles for
      accelerating succession and evolution 50
      attitudinal principles 54
      biological resources 48-49
      catch and store energy 49-50
      creatively responding to change 54-55
      diversity 50-51
      edge effects 52
      efficient energy planning 48
      energy cycling 49-50
      functions 48
      landscape 54
      observe and interact 47-48
      problem and its solution 54
      relative location 48

self regulation and feedback 50
small-scale intensive systems 50
through patterns 52-53
waste production 53-54
yield 54
Douglas fir 79, 82-84, 119
Dragonflies 89, 112
Drengson, Alan 28
Duchy of Cornwall 118
Ducks 16, 89, 112
Dunbeag Project 97, 98-101
Dwarf muntjac 140

Edge effects 52
Employment 101
Energy cycling 49-50
English Nature (EN) 88, 145, 152
English Woodland Grant Scheme (EWGS) 37, 100, 109, 118, 144, 145, 155
Environmental Stewardship Grants 155
EU Habitats Directive 146
European Court of Justice (ECJ) 146
European larch 16, 37, 79, 83, 99
Extraction 2, 14, 36-37, 48, 80-81, 83, 84, 92, 94-95, 100, 104, 105, 113

Fairlie, Simon 154, 178
Farm Woodland Premium Scheme 155
Felling licences 37, 144-145, 155
Fencing 17, 40, 42, 49, 50, 59, 62, 63, 73, 74, 85, 101, 102, 113, 155
  garden 119
Ferns 5, 14, 50, 109
Fertiliser 9, 16, 49, 50, 61
Financial capital 25, 192
Food distribution system 160
Forest dwellers
  non-timber products, use of 122
  patterns of 28
  sustainable rural livelihoods 21-27
Forest garden 85, 122, 134
Forest management 16, 79, 81, 84, 118
Forestry Commission 3, 13, 14, 37, 63, 66, 74, 79, 86, 100, 104, 105, 118, 144, 146, 147, 155, 161
Forestry Contracting Association 146, 155

Forestry Enterprise 3, 150, 151
Forest Stewardship Council (FSC) 107, 118, 119
Fossil fuels 21, 29, 55, 105, 119, 158, 160
Fruit harvesting 23
Fungi 4, 23, 52, 53, 88, 122, 137-138

General Permitted Development Order (1995) 148, 152
Glenelg 98, 99
Goats 11, 62
Gorse 16, 46, 52, 129
Goudhurst 55-57
Grafting 126
  cleft grafting (top-working) 132, 134
Grand fir 99
Grants 155
Grazing 2, 6, 12, 16, 23, 59, 74, 81
Green manures 49, 50, 61
Green wood 95-96, 98, 108
Ground cover, in woodland 66, 131
Group felling 81

Hardwoods 83-84, 138
  air drying 98
Hart, Cyril 65
Hart, Robert 123
Hawthorn 57, 60, 66, 98, 127, 132, 168
Hazel underwood 7, 37, 126-127
Health and Safety 146
Heartwood Forest initiative 67
Hedgerow Act (1997) 17
Hedgerows 16-17, 46, 52, 147
Herbaceous plants 50, 78, 130
High forest management
  felling of high forest
    clear felling 80-81
    converting plantations into diverse woodland 81-83
    group felling 81
    selective felling 81
    thinning 79-80
  High forest produce, marketing of 117
Hiley, W.E. 28
Himalayan balsalm 66
Holmgren, David 47
Honey 57, 111, 123, 139, 140, 160
Horse
  Clydesdale 95
  Suffolk 95

Horse extraction 36, 37, 81, 94
Human capital 25
Humus 4

Indian oyster mushrooms 99
Indonesian rainforests 21, 105, 158
Insurance 2, 146
Integrated forestry agriculture system 122
Interdepartmental Committee on the New Development of Contaminated Land 66
International Federation of Organic Agricultural Movements (IFOAM) 118
Invasive species 65-66
Invertebrates 53, 59, 88
Iron horse 84, 94
Irregular forest management 16

Japanese knotweed 66
Joinery 13, 63, 108-109
Juglans regia 126

Kestrel 89
Keveral Farm 13
Kilfinan Community Forest 100-101
Kiln drying 97
Kindling 66, 71, 102, 107, 111
Kingfisher 89, 113

Laetiporus sulphureus 137
The Land Is Ours 31
Larch Piece 42
Laurel 65
Layering 40-41, 78
Leckmelm wood 83, 86
Leguminous plants 48, 50, 66
Lichens 14, 50, 88, 109
Lime 2, 57, 66, 84, 108, 139
Limestone scalpings 37
Lion 75
Livestock 10, 12, 17, 46, 49, 50, 63
Lizards 106, 112, 145
Local Exchange Trading System (LETS) 25
Local Government Ombudsman 154
Localisation, importance of 119, 160
Local provenance 59, 60
Lodgepole pine 79, 83, 99
Lodsworth 77, 109, 125, 127, 160
Lupins 61

Maidstone chestnut coppice 101
Manures 49, 50
  animal 49
  bird 3, 4, 52
  green 49, 50, 61
Maples 23, 42, 66, 102, 113
Marketing 117
Meat 140
  wild boar 141
Medicinal herbs 113, 137
Meera's wood 58-59
Microclimate 36
Milling 92-93
Milne, A. A. 20
Ministry of Agriculture, Fisheries and Food (MAFF) 16
Mixed plantations 16
Modern foresters 14
  patterns of 28
Mollison, Bill 47
Monkey puzzle tree 134
Monocultural plantations 5, 51, 81
Mosses 5, 14, 50, 76, 88, 109
Mulching 59, 60
Mushrooms 15, 40, 42, 99, 137, 138

National Forest Initiative 67
National Planning Policy Framework (NPPF) 147, 148, 150
National Proficiency Tests Council 147
National Small Woods Association 3
National Vegetation Classification (NVC) 14
Natural capital 25
Natural England (NE) 145
Natural regeneration 2, 16, 46-50, 65, 80-82, 99-100, 141
Nature Conservation Body 145
Nesting boxes 89
Nightjars 6, 53, 111, 112
Nitrogen fixation 4, 12, 66
Non-residential building 150
Norway spruce 16, 79, 99
Notch planting 59, 66
Nurse crop 16, 17, 79
Nuthatch 88, 113

Oak 2, 4, 7, 8, 16, 37, 40, 57, 58, 60, 79, 98, 100, 108, 113
  cork 12, 13
  holm 12, 124
  timber framing 119, 147

Index 189

Oak leaf 126
Orchards 16, 29, 51, 65, 67, 85, 132, 145
Orchid 23
   early purple 23
Orchis mascula 23, 37
Organic food 117, 118
Organic wastes 113
Owls 112, 113
Oyster mushrooms 99, 138

Pacific Certification Council (PCC) 119
Permaculture 25, 47, 49-51, 54, 55, 99, 122, 147
Pheasants 65, 140
Physical capital 25
Phytophora 66
Pigeons 4, 140
Pigs 12, 48, 99, 150
Pit planting 66
Pit sawing 92
Planning Guidance Note for Rural Workers 150
Planning Inspectorate 148
Planning law
   appeal process 148-150
   application 148
      for forestry dwelling 150-152
      for non-residential building 150
   brief history of 147
   permission 148
   prior notification procedure 150
   reference documents 148
   temporary permitted stays on forestry land
      28 Day Rule 152
      seasonal forestry workers 152-153
Planning Policy Statements (PPSs) 150
Plantations
   broadleaf 14, 16
   concept of 12-13
   coniferous 14-15, 34, 63, 79, 86, 89, 93, 98, 101
   mixed 16
   monocultural 5, 51, 81
Planting 134
Planting initiatives 67
Planting trees 59-60
   aftercare 63
   bare root/container grown trees 62
   beating up 63
   on degraded land 66-67
   green manures in nursery beds 61-62
   invasive species 65-66
   planting patterns 63
   plant spacing 63
   pruning 63-65
   stratification 62
   tree nursery 60-61
   tree shelters 62-63
Plant spacing 63
Plastics 62, 74, 107
Pleachers 17
Pollarding 10, 11
Pollen analysis 2
Polytunnel 85, 97, 99
Poultry 50
Prickly Nut Wood 23, 25, 37, 49, 50, 52, 54, 109-116, 126, 130, 152
   birds 113
   diversity of activities carried out at 24
   longer-term objectives 113
   management objectives 109-111
   planning experience at 153-154
Programme for the Endorsement of Forest Certification (PEFC) 118
Pruning 63-65, 126, 132
Pulpmills 117
Pussy willow 139

Quercus robur 57, 58, 124, 169
Quince 134

Rabbits 2, 36, 42, 46, 60, 62, 63, 71, 73-76, 140
Rainfall 4, 35, 36, 62, 83
Rainwater harvesting 48, 160
Reforestation 66
Reforesting Scotland 6
Renewable Heat Incentive 104
Renewable wood resource 104
Rhododendron 39, 41, 65-66, 111, 113, 155
Ride management 89-90
Rights of Way Act (2000) 145
Root nodules 4
Rootstocks 16, 132, 134, 200
Roundwood timber 54, 113, 119
Roundwood Timber Framing Company 113
Royal Forestry Society 155
Run-off 3, 4
Rural economy 31, 182
Rural employment 3, 119, 160
Rural livelihoods, sustainable 22-27
Rural sustainability 111

Salix caprea 139
Sandstone hoggin 37
Sawdust 92, 138
Sawlogs 37, 117, 138
Sawmilling 37, 99
Sawmills 29, 49, 80, 81, 92-93, 95, 98, 113, 117, 146, 180
Scotland Rural Development Programme (SRDP) 85
Scots pine 2, 16, 37, 62, 79, 83, 109
Scottish Natural Heritage (SNH) 145
Scottish Skills Testing Service 147
Sea evaporation 4
Seasoning wood 96-98
Seed, germination of 4
Selective felling 14, 81
SGS Forestry in Great Britain 118
Shelter belts 3-4, 12, 54
'Shelterwood' system of management 16, 82
Shield budding 132, 134
Shredding 10-11, 49
Shrinkage 96
Shrubs 14, 50, 129
Singling 7-8, 39
Site of Special Scientific Interest (SSSI) 37, 50, 145-146, 152
Sitka spruce 14, 79, 83, 144
Sledge 94
Snigging 94, 95
Social capital 25
Softwood 80, 119
Soil
   acidification. See Acidic soil
   enrichment 4
   erosion 4, 6, 14, 17, 54, 66, 119
   micro-organisms 3, 4
   stabilisation 66
Soil Association 117, 118
Squirrels 4, 36, 42, 73-76, 126, 140
Steam bending 107-108
Stooling 78
Stools 6, 7, 8, 40-42, 65, 71, 73, 76, 78, 132, 163
Subsoil 3, 4, 59, 66
Suckering 2, 7
Suffolk 7, 95
Sun traps 52
Surrey Coppice Group 25, 155
Sussex Coppice Group 25, 155
Sussex Downs Conservation Board 152
Sustainable development 31, 119, 160, 182
Swales 48, 58
Symbiotic relationships 3, 4, 6, 63, 138

Taxation 147
Taylor, Duncan 28
Tenax 74
Thatching spars 10, 70, 168, 170
Thinning 14, 16, 42, 49, 79-80, 82
Tighnabruaigh 97, 98
Timber 2, 13, 23, 51
   roundwood 54, 113, 119
Timber arch 94
Timber frame 29, 31, 54, 81, 113, 119, 147
'Timber miles' campaign 158
Time stacking 50
Tinkers Bubble 29
Town and Country Planning Act (1947) 147, 178
Transpiration 3-5
Tree Council 155
Tree nursery 60-61, 180
Tree Preservation Order 145, 161
Trees For Life 6
Tree shelters 59, 62, 63
Trefoil 61, 62, 66
Tuber uncinatum 138

UK Woodland Assurance Standard (UKWAS) 118, 155
Ullapool 83
Underwood 98, 101, 104, 109, 145
University of Surrey Civil Engineering Department 119

Veneer 15, 79, 117, 167, 170
Ventilation 62, 180
Veteran trees 88
Violets 34, 87, 88

Walnut 17, 25, 50, 126, 134, 163
Walnut leaf wine 126
Water
   infiltration 66
   in woodlands 89
Weaving 108

Weed species 14, 82
Wessex Coppice Group 70, 71
Western hemlock 82, 99
Western red cedar 82
Whip and tongue grafting 132-133
Wild boar 141
Wildlife and Countryside Act (1981) 145
Wildlife corridor 17, 48, 54, 109, 160, 183
Wildling (pip grown) apples 132
Wildwood 2, 122, 147
Wood buildings 31
Woodfuel 104-105
   sustainable 105
Woodland Assessment Grant (WAG) 155
Woodland community
   low impact development 30-31
   re-emerging 28-30
   residential site 30
Woodland flora, assessment of 36-37
Woodland Improvement Grants (WIG) 66, 70, 74, 104, 105, 111, 155

Woodland management 3
   coppicing 70
   deer, rabbits, squirrels and coppice 73-76
   differences with present forestry practice 159
   future of 158
   initiatives in 119
   layering 78
   observation and recording 34-36
   old records 34
   overstood, stored or neglected coppice 76-78
   recommendations 160-161
   stooling 78
   for wildlife 86-89
      coppicing 87-88
      dead wood 88
      minimum intervention 87
      nesting boxes 89
      ride management 88-89
      veteran trees 88
      water 89
   woodland flora, assessment of 36-37
      first assessment 37
      notes from the walk 39-43

Woodland Management Grant (WMG) 155
Woodland Planning Grant (WPG) 155
Woodland Regeneration Grant (WRG) 155
Woodlands
   conservation projects 107
   ecosystem 3-5, 53
   habitats 2
   regeneration of
      lobular pattern 52
      natural process 46
      planting 46-47
   resources 3, 111, 158
   sustainable development in 119
   and taxation 147
   types of
      ancient woodland 5
      broadleaf plantations 14
      Caledonian pinewoods 6
      coniferous plantations 14-16
      continuous cover forestry 16
      coppice 6-7
      hedgerows 16-17

      mixed plantations 16
      National Vegetation Classification (NVC) of 14
      orchards 16
      overstood, stored or neglected coppice 8
      plantations 12-13
      pollarding 10
      shelter belts 12
      short rotation coppice 9
      shredding 10-11
      suckering 7
      wood pasture 11-12
Woodland survey 35-36
Woodland Trust 3, 67
Woodmill 97
Wood pasture 2, 11-12, 16, 34, 160
Wood stations 158, 160
Working Horse Trust 95
Wych elm 7, 84, 108

Yanesha Indian co-operative 82
Yew 5, 62, 108
Yurt 102, 107, 112, 116, 153, 154, 167

Also by Ben Law

THE WOODLAND HOUSE
In 2003 Ben Law captivated the nation by building his woodland house on Channel 4's Grand Designs programme. Full of stunning colour photographs, this is a visual guide to how Ben built his outstandingly beautiful home in the woods. It is also a practical manual and the story of a man realising a lifetime's dream to build one of the most sustainable and beautiful homes in Britain. 96pp. Pbk. £14.95

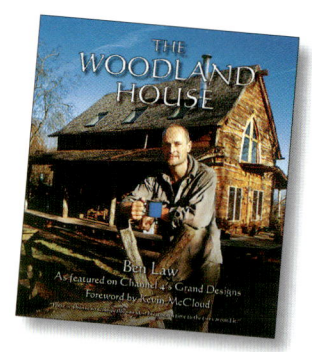

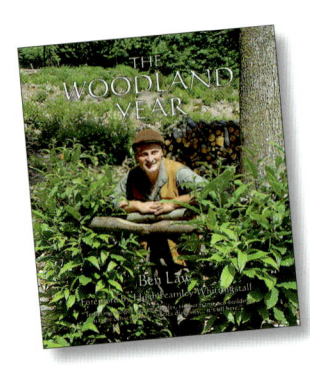

THE WOODLAND YEAR
Packed with beautiful colour photographs, this is an intimate month-by-month journey through Ben's yearly cycle of work, his naturally attuned lifestyle and his deep understanding of the wood in which he lives. Each month includes guest contributions from woodlanders in other parts of England and Wales. 176pp. Hbk. £19.95

ROUNDWOOD TIMBER FRAMING
Filled with detailed colour photographs and drawings, this unique and practical 'how to' book, is unquestionably a benchmark for sustainable building. Roundwood Timber Framing encourages communication between woodsmen, planners, architects and builders and helps to close the loop between environmental conservation, use of renewable local resources and the regeneration and evolution of traditional skills to create durable, ecological and beautiful buildings. 168pp. Hbk. £18.95

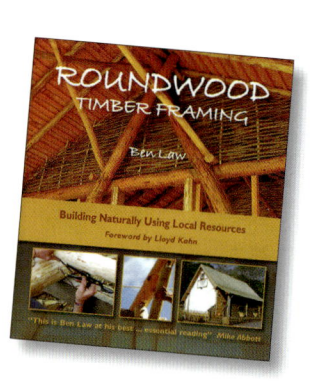

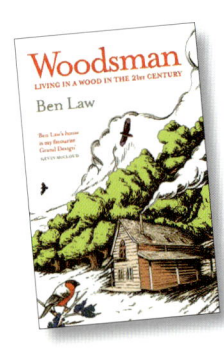

WOODSMAN
Living in a Wood in the 21st Century
This is the story of Ben Law, the woodsman. It takes us through the 20 years of his life in Prickly Nut Woods; coppicing, building and living. But this is also the tale of the wood itself, how it lives and breathes and how it has developed alongside Ben and his woodsman's life. 256pp. Hbk. £14.95

ALL THESE & MORE
AVAILABLE FROM ALL GOOD STORES IN THE ENGLISH SPEAKING WORLD
AND FROM:

www.permanentpublications.co.uk

More books from Permanent Publications

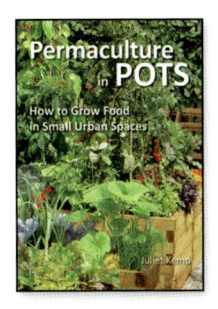

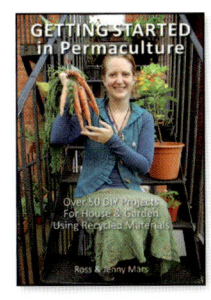

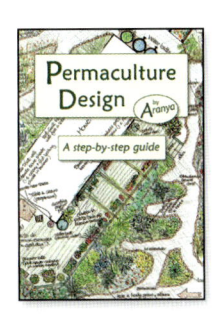

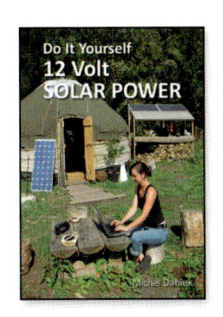

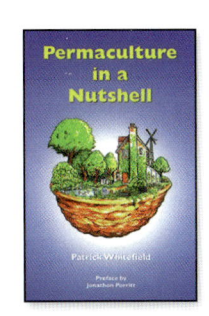

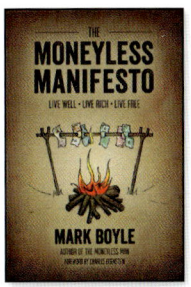

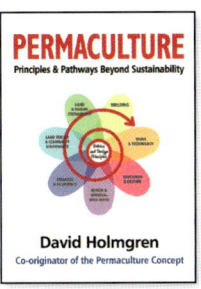

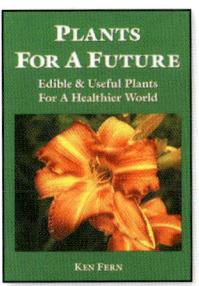

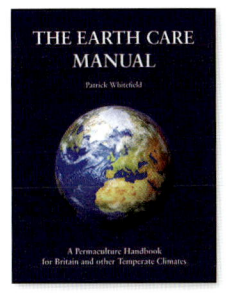

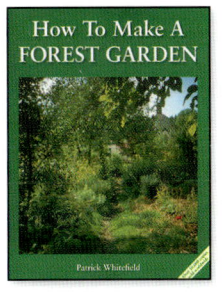

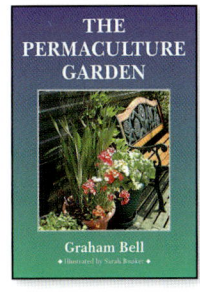

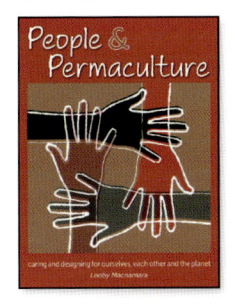

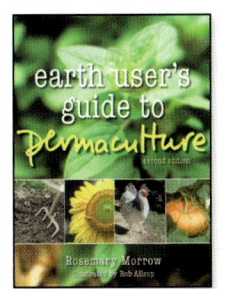

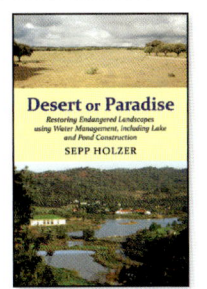

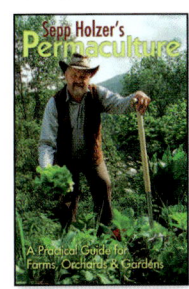

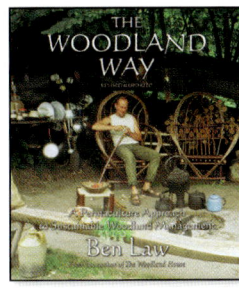

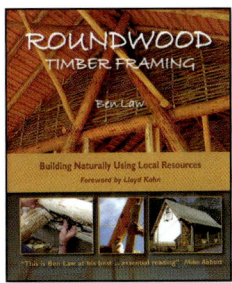

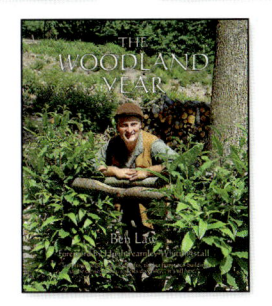

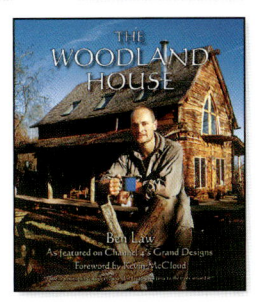

ALL THESE & MORE
AVAILABLE FROM ALL GOOD STORES IN THE ENGLISH SPEAKING WORLD
AND FROM:

www.permanentpublications.co.uk

Inspiration for Sustainable Living

Permaculture is a magazine that helps you transform your home, garden and community, and save you money.

Permaculture magazine offers tried and tested ways of creating flexible, low cost approaches to sustainable living, helping you to:

- Make informed ethical choices
- Grow and source organic food
- Put more into your local community
- Build energy efficiency into your home
- Find courses, contacts and opportunities
- Live in harmony with people and the planet

Permaculture magazine is published quarterly for enquiring minds and original thinkers everywhere. Each issue gives you practical, thought provoking articles written by leading experts as well as fantastic ecofriendly tips from readers!

permaculture, ecovillages, ecobuilding, organic gardening, agroforestry, sustainable agriculture, appropriate technology, downshifting, community development, human-scale economy ... and much more!

Permaculture magazine gives you access to a unique network of people and introduces you to pioneering projects in Britain and around the world.
Subscribe today and start enriching your life without overburdening the planet!

PERMANENT PUBLICATIONS
The Sustainability Centre, East Meon, Hampshire GU32 1HR, UK
Tel: 01730 823 311 Fax: 01730 823 322 (Overseas: int code +44-1730)
Email: info@permaculture.co.uk

To subscribe and for daily updates, vist our exciting and dynamic website:
www.permaculture.co.uk